Images of Time

The Role of a Historical Consciousness of Time in Learning History

Images of Time

The Role of a Historical Consciousness of Time in Learning History

Arie Wilschut

Amsterdam University of Applied Sciences

IAP

INFORMATION AGE PUBLISHING, INC.
Charlotte, NC • www.infoagepub.com

Library of Congress Cataloging-in-Publication Data

Wilschut, Arie.
[Beelden van tijd. English]
Images of time : the role of a historical consciousness of time in learning history / Arie Wilschut.
p. cm.
Translated from Dutch.
Includes bibliographical references.
ISBN 978-1-61735-906-4 (pbk.) – ISBN 978-1-61735-907-1 (hardcover) – ISBN 978-1-61735-908-8 (ebook) 1. History–Study and teaching–Philosophy. 2. Time–Philosophy. I. Title.
D16.2.W5513 2012
907–dc23

2012020702

Translated from Dutch
Original title
Beelden van tijd: de rol van historisch tijdsbewustzijn bij het leren van geschiedenis
Assen, Netherlands: Van Gorcum (2011)
ISBN 9 789023 248118

This translation has been made possible by a grant from the Department of Education of the Amsterdam University of Applied Sciences (Hogeschool van Amsterdam).

Printed and bound by CPI Group (UK) Ltd, Croydon, CR0 4YY

Contents

Preface ..... **vii**

Introduction ..... **ix**

1 The Development of History Education **1**

Developments Up to the 1960s 2

The Crisis of the 1960s and the Responses to It 10

New Socio-Political Interest in History Education from the 1990s Onwards 21

Conclusions 28

2 Historical Time **33**

Objective Time 34

Subjective Time 41

Historical Time 47

Alternatives to a Historical Consciousness of Time 71

Historical Time Consciousness in History Education: Aims and Research Questions 79

Retrospective and Preview 90

3 Developing and Learning a Consciousness of Time: Review of Current Research **95**

Results from Developmental Psychology Research 96

Domain-Specific and General Educational Research Up to the 1960s 105

Domain-Specific and General Educational Research After the 1960s 114

Diverse Data from Educational Research About the Six Key Concepts of a Historical Consciousness of Time 121
Conclusions and Consequences for Future Research 139

4 Empirical Research into Students' Orientation in Historical Time 143
Research Design 145
The Test Used During the Pre-Test and Post-Test 161
Results 166
Discussion of Results 174
Conclusions 179

5 Conclusion 183
Consequences for Teaching 188
Significance for Society 190

Notes 193

References 197

Appendices

A PowerPoint Presentations Used in the Lessons According to the Chronological and Imaginative-Associative Approaches 217

B Content Protocols for the Lessons According to the Chronological and Imaginative-Associative Approaches 231

C Summaries of the Subject Material Handed Out to the Students 255

D Questions Used for the Pre-Test and Post-Test 263

E Rules for Scoring the Pre-Test and Post-Test 273

F Average Scores to All 25 Items During the Pre-Test and Post-Test 279

G Results of the Paired t-Tests Applied to the Differences Between Item Scores on the Pre-Test and Post-Test in Group A and Group B 281

H Significance of the Difference between Groups A and B as Summarized in Table 4.11 283

About the Author 285

Preface

It has become commonplace in the field of teaching and learning history to state that the subject is about far more than just memorizing facts about the past. Historical thinking, historical reasoning, historical understanding, or the "meta-concepts" of history are the mainstream in research on teaching and learning history. A more or less standard set of concepts has been developed to characterize this aspect of the subject: concepts like evidence, interpretation and bias, causation, continuity and change. However, not all of these seem to belong to one and the same category. After all, interpreting evidence and establishing facts based on reliable, representative, and unbiased information is the concern of all academic disciplines, and so is considering the causes and consequences of events and developments. Comparing different periods in time to find out to what extent there has been continuity and change, and trying to take the perspective of past times seriously by "standing in other people's shoes," on the other hand, seem to be typical of history, because the viewpoints of different times are at stake. This consideration induced me to wonder whether or not it is the phenomenon of time that is the essence of history, and therefore learning to think in terms of time its main educational objective. Scrutinizing evidence on its reliability and representativity is one thing, but putting the material in the correct context of its time is another. Studying causes and consequences of developments in human society is a complicated matter, but even more so when it comes to causes and consequences of events and developments in circumstances fundamentally different from those of the present.

Images of Time, pages vii–viii

I am therefore grateful that my employer, the Department of Education in the Amsterdam University of Applied Sciences, has given me ample opportunity to study this matter in depth, which resulted in a book published in Dutch in June, 2011. Subsequently my employer also provided the funds to bring about this English translation of the book. In this way, I hope to have been able to add some new perspectives to the research into teaching and learning history, which will hopefully benefit teachers of history not only in my own country, but also throughout the world.

—Arie Wilschut

Introduction

This is a book about teaching and learning history: a subject field bordering different worlds—the world of teaching and learning and the world of history.[1] It goes without saying that educational insights play a major role in the design of learning environments for the school subject of history. However, expertise in history is also of great importance, since an effective curriculum design requires clarity in terms of the nature of the knowledge and the modes of thinking that students should acquire. This is why primarily expertise in the theory of history is helpful. The objectives of education and the associated instructional formats are for the main determined by the nature of the subject taught.

Expertise in teaching and learning history is therefore a hybrid branch of academic studies. There is a great distance between the cultures of the subject areas facing the domain-specific educationalist. Educational science is continuing to develop in the direction of an empirical social science, where measurement leads to knowledge, and in which attempts are made to come to grips with the complicated domain of thinking and learning processes through statistical calculations. The theory of history, on the other hand, is based on philosophical reflection that, at first glance, hardly gives the impression of being an exact science. Interpretative conceptualization in the form of historical narratives also plays a major role in history itself. No wonder that there are enormous differences in the appearance of publications in the two subject areas: the educationalist prefers to publish articles, drafted according to a standard template and including tables and

Images of Time, pages ix–xvii

graphs; philosophers of history and historians prefer to produce hefty books in which linguistic tools and images help express certain trains of thought.

Experts in teaching and learning history are either historians who have acquired some educational knowledge, or educationalists who have looked a little into the principles of historical thinking. They have no other option, for—in the Netherlands and most other countries—teaching and learning history is not taught as an academic subject. There are no training programs for experts in teaching and learning history. To a certain extent, therefore, an expert in teaching and learning history is always a "self-made" man or woman. Obviously his or her academic origins in one of the rather distant worlds of history or educational science will always be recognizable in his or her work to a certain degree. The author of this book has been trained as a historian, and is therefore an autodidact when it comes to educational science. Naturally, this has affected the choices that were made in this book and the views developed.

The gray area between the worlds of the historical and educational sciences may be regarded as somewhat contested. Which of the two worlds has the right to dominate the other? Should expertise in teaching and learning history be viewed as a branch of educational science, which merely has to apply insights from that science to the school subject? Or is it a branch of academic history, which is merely tasked with shaping the academic discipline into a particular form that is comprehensible to students?

Examples are available for both sets of views. For instance, in the 1990s, the Dutch educationalists Lowyck and Verloop published their handbook entitled *Onderwijskunde* [Educational Science], which included a series of "subject-specific translations" in which general educational insights were applied to school subjects (Lowyck & Verloop, 1995; Dalhuisen & Walhain, 1997). The handbook was intended for teacher training, in which both general and domain-specific educational science have always been fixed elements. An attempt towards harmonization in that area is understandable, and in this case implied a positioning of general educational science as the core science to which the domain-specific teaching and learning experts had to adapt. This view is also evident in examples of scientific research. In a study by Perfetti, Britt, and Georgi (1995) into text-based learning, for instance, the subject of history does not originally appear to have been an important starting point. In actual fact, the authors were only interested in the emergent learning processes founded on the texts:

> To some extent, we use history merely as an interesting subject-matter window on these cognitive problems. Ours is not a contribution to how students acquire knowledge in the discipline of history, although we do discuss some

> comparisons of history 'experts' and 'novices.' (Perfetti, Britt, & Georgi, 1995, p. vii)

Recent Dutch dissertations display something similar. The doctoral thesis by Van Drie (2005), for example, revolved around the application of the principles of computer-supported collaborative learning (CSCL) to teaching history, and Prangsma's thesis (2007) focused on the application of general educational insights from the domain of multimodal representation in collaborative learning. The manner in which students learn history was the research topic in all these studies. The research questions associated with this, however, were generated by general educational science.

The project *How People Learn* by the American National Academy of Sciences was, at first, a general educational scientific enterprise. Attempts were made in an initial study to collate insights gained from empirical research into the ways people learn, in order to arrive at the—"evidence based"—best possible methods of educating (Bransford, Brown, & Cocking, 1999). These general insights were subsequently applied to the school subjects of history, mathematics, and sciences (Donovan & Bransford, 2005). However, one of the general insights determined based on empirical studies was that:

> Different domains of knowledge, such as science, mathematics, and history, have different organizing properties. It follows, therefore, that to have an in-depth grasp of an area requires knowledge about both the content of the subject and the broader structural organization of the subject. (Bransford et al., 1999, p. 226)

No wonder, then, that further detailed elaboration of the general principles of the project created a great amount of space for a subject-specific approach. For instance, in the section focused on history, a great deal of attention was paid to the British research by Lee and Ashby (Lee, 1983, 2004; Lee & Ashby, 2000, 2001) and an American study by Wineburg (2001), who based the design of their research questions for empirical investigations on history-specific modes of thought. The focus on the domain-specific aspect of learning processes, including the subject-specific organizational structures of knowledge and modes of thought, is increasing. The American educationalist, Shulman, for instance, argues for what he calls "pedagogical content knowledge": a teacher should possess a profound knowledge of his subject and modes of representation associated with it that could serve to clarify something to someone else—the strongest analogies, illustrations, examples, ways of explaining, and demonstrations. He should also know what makes a subject difficult to learn and which misconceptions may arise among students in the context of a certain subject (Shulman, 2004).

The specific nature of the subject area inevitably demands the attention of the educationalist specializing in teaching and learning history. Van Drie (2005), for example, asked herself what historical reasoning actually constitutes, before measuring the effect of a certain educational approach on that reasoning. Perfetti et al. (1995) started off with a description of historical reasoning as constituting nothing more than reasoning in general, but then stumbled across the fact that it is a manner of dealing with texts which has some specific features, which make it "neither specifically historical, nor fully general" (p. 5). The school subject of history could not be disregarded as experts and novices became involved into their research. This empirical approach to domain-specific educational studies, also applied by Wineburg (1991), Baker (1994), Greene (1994), and Van Boxtel & Van Drie (2004), indirectly considers the content and structure of a subject domain, because it implicitly assumes that the experts (for instance teachers or academics) have a better command of a subject than novices (students), and that the experts can therefore point out the direction in which the students should develop. Thus, the specific nature and structure of a discipline is brought into focus via the contribution of the experts.

A more direct focus on the nature and structure of a subject is also possible. For instance, one may choose studies in the area of historical theory as a starting point for the development of research questions concerning the teaching and learning of history. Lee (1983) noted that there are good reasons to relate teaching and learning history to the philosophy or theory of history: "first, in suggesting what the structure of the discipline might be; and second, in providing an analytical basis for investigation of the 'thinking tasks' confronting children" (p. 20). German expertise in teaching and learning history has particularly emphasized the importance of historical theory. A focus on the theory of history is necessary, according to German authors in the 1970s, because expertise in teaching and learning history can simply not do without: "Expertise in teaching and learning history (*Geschichtsdidaktik*) cannot by itself design a system of categories, it has to rely on the theory of history in this matter" (Mayer & Pandel, 1976, p. 31, my transl.). "Expertise in teaching and learning history has to rely on the results of historical-theoretical analyses for the completion of its tasks; without these analyses it would not be in a position to make sufficiently motivated decisions in matters concerning the nature of historical knowledge" (Rüsen, 1975, p. 436, my transl.). A great deal of research within German expertise in teaching and learning history has been conducted—based on these principles—into the nature, appearances, and roles of *Geschichtsbewusstein*, which is probably best translated as "historical awareness."

According to Jeissmann (2000), the academic discipline developing expertise in teaching and learning history has three tasks: firstly, to investigate in which contents and modes historical awareness appears in society; secondly, to look into the functions of such historical awareness for the self-concept of a society; and thirdly, to develop methods for the teaching of history based on theoretical insights into the structure of historical knowledge and in connection with historical awareness and its functions in society. In this way, the discipline is assigned an elaborate task: it develops into a *Wissenschaft vom Geschichtsbewusstsein in der Gesellschaft,* a science of historical awareness in society. As a matter of fact German experts in teaching and learning history have published numerous theoretical, educational and empirical studies into the ways of appearance of historical awareness in society and among students. A number of studies conducted in Germany has been collated in Von Borries, Pandel, & Rüsen's 1991 book. A large scale survey was carried out in Europe among 32,000 students in 27 countries in the project *Youth and History* (Angvik & Von Borries, 1997). These studies have, however, not rendered methods that can be directly applied in daily school practice. The lack of practice-oriented empirical research has been noted in Germany since the start of the new millennium:

> Expertise in teaching and learning history can no longer be satisfied with impressive sounding great concepts like reflective historical awareness and historical thinking. These concepts, which deal with the fundamental structures of historical understanding, must be put into operation, must be made fit to the size of processes of teaching and learning, in such a way that they can be maintained in the practice of teaching and assessed by empirical research. (Günther-Arndt & Sauer, 2006, p. 8–9, my transl.)

An example of this approach in the field of teaching and learning history in the German-speaking area has come from Switzerland; through an extensive study of lessons delivered by experienced history teachers, attempts were made to map out the principles of good history education (Gautschi, 2009).

The diverse directions into which expertise in teaching and learning history may develop is clearly reflected in the modalities of the German expertise in teaching and learning history: the direction of theory-based study, leading to definitions of historical awareness and subsequent empirical research into its appearances, and the direction that chooses the empirical study of educational and learning processes in schools as its starting point. In 1981, the Dutch expert in teaching and learning history Toebes, who was then one of the first in the Netherlands to obtain a doctoral degree based on a thesis on teaching and learning history, claimed that it was a misconception to regard domain-specific educational science as no more than the application of general educational science to the respective school subjects. It would be just as incorrect, according to Toebes, to view domain-specific educational science as an insignificant appendage to the academic discipline related to the school subject concerned. Domain-specific educational science takes its independent position between the two: "Domain-specific educational science is characterised by its position between the academic discipline (in this case, academic history), general educational science and society, in regard to which it has a coordinating and integrating role" (Toebes, 1981, p. 5). The above-mentioned exploration of the modalities in which the expertise in teaching and learning history has developed shows to what extent Toebes was right. Neither educational science nor academic history should dominate; expertise in teaching and learning history should find its own place in the academic spectrum, which should not be too theoretical, but not too exclusively empirical either.

Besides history and educational science, Toebes submits a third determining factor, *society*, by order of which education takes place. There is a justification for referring to this factor as an element of expertise in teaching and learning history, even if there is no intention to promote it, according to the broad German definition, to a "study of historical awareness in society." Society has placed history as a subject on the school curriculum by means of political decisions, as it has certain expectations of the teaching of that subject. These expectations exercise an influence on teaching objectives and therefore also on the choice of teaching methods associated with these aims. One could therefore also argue that the relationships among society, politics and history teaching are *not* an appropriate theme for expertise in teaching and learning history, as society and politics simply have a decisive role in terms of the contents and aims of education: the history teacher simply has to accept these choices and to use his/her combination of subject expertise and educational knowledge to find the most appropriate teaching methods to realize certain learning objectives set by others. However, this summary of the state of affairs is perhaps a little too simplistic. It is not always certain that society's expectations that come to the fore in public debates can be honored by history teaching, given the nature of academic history and historical thought.

The relationship between historical awareness as present in society and developments in history teaching is complex and deserves an explicit focus—as has been emphasized in German research—although education needs to retain its central place within this, and the study of historical awareness should not become a goal of its own. It is desirable that experts join in the debate about educational objectives for history education, as highlighted for instance in Grever, Jonker, Ribbens & Stuurman (2006);

Grever & Stuurman (2007); and Symcox & Wilschut (2009). The study of historical developments in history curricula may be useful in this respect in order to place current developments into perspective, as has been demonstrated in a number of studies in this area (Toebes, 1981; Nash, Crabtree & Dunn, 1997; Phillips, 1998; Symcox, 2002).

The above arguments suggest that an attempt at defining the academic discipline that creates expertise in teaching and learning history may be made as follows. Historical theory provides an insight into the distinctive nature of knowledge and ways of thinking, which are the object of the school subject and therefore, in that sense, a guide for questions and problems facing teachers. The relationship between society and politics is important because both have expectations and place demands on education; these need to be reviewed to determine whether they are correct from a historical point of view and to find out how they could be realized. Educational science throws light on the methods that can be successfully applied in teaching practice. These three components together characterize the discipline that creates expertise in teaching and learning history as it is described in this book.

Research into teaching and learning history aims at theoretical studies concerning history theory and philosophy (which focus on the specific knowledge structures and modes of thought that could be transferred into educational objectives), at historical studies of developments in history education and the history curriculum (with a focus on the relationship between society and education), and finally at empirical studies into the application of teaching methods in teaching practice. This book will focus on each of these three aspects of expertise in teaching and learning history. The advantage of this is that it creates an integral perspective around the kind of expertise in teaching and learning history that will be covered in this book. The disadvantage, however, is that none of the three areas can be studied in depth within a study of a relatively limited size. This book may therefore be viewed as a basis for a series of follow-up studies.

Taking the consideration that domain-specific modes of thought are a major guiding principle for domain-specific educational science as a point of departure, this study has been directed at the topic of "historical time" as its central focus. This was chosen based on the consideration that reasoning in terms of the passage of time is one of the most characteristic aspects of history: "History is the systematic study of the past, and at its heart is time" (Jordanova, 2006, p. 105). Organizing knowledge in a way that takes into account the passage of time and reasoning in terms of the passage of time are characteristic features of learning history. The question facing the expert in teaching and learning history that follows from this is: which are the learning problems associated with an awareness of historical time? There is some justification for the assumption that certain modes of thought about time have to be acquired in history education, given the condition that history uses time dimensions that are of an entirely different order of magnitude than the daily awareness of time which each person develops spontaneously. History is not about yesterday, today and tomorrow, and very often it is not about the passage of time amounting to a single human life; rather it is about time spans of hundreds or thousands of years, which are difficult for people to imagine. The central question in this study is therefore:

> What role does an awareness of historical time play in learning history?

In order to be able to answer this question, the precise nature of historical time has been explored as well as to what degree this awareness of time differs from other images and experiences of time. Use has been made of philosophical, psychological and anthropological studies, as well as studies into theory of history. From the definition of the concept of "historical time," which has been developed in this manner, a number of central categories have been derived meant for use in a kind of history education that takes its point of departure in awareness of historical time as a fundamental component. Based on these central categories, a number of research questions have been developed that can provide the basis for empirical educational research, including the development of a research agenda for expertise in history education. An empirical study has been conducted to address a number of questions detailed on this agenda, the results of which are reported in this book. The book is opened by an introductory chapter containing a brief historical overview of the developments in history education, in order to determine to what extent the expertise on teaching and learning history developed in this book represents a renewal of or supplement to the discipline as it currently stands. This will also answer the question of what role an awareness of historical time has played thus far in the objectives of history teaching.

The book therefore has been drawn up around the following questions:

1. How have history education and educational expertise in teaching and learning history developed thus far, and what role has the awareness of historical time played in this?
2. What is historical time, and how is this experience of time distinct from other images and experiences of time?
3. What important categories for teaching and learning history can be derived from the definition of historical time?

4. What is known up to now from psychological and educational research about the ways in which students orientate themselves in historical time?
5. What are the optimum methods to promote orientation in historical time with students?

These questions relate to the three subfields of expertise in teaching and learning history that have been set out in this introduction: the relationship among history education, society and politics (the first question); historical theory and philosophical studies that may shed light on educational objectives (the second and third questions); and empirical studies into the effect of teaching methods (the fourth and final question). Answering these questions coherently will provide a perspective on history education that will transcend the contrast frequently occurring in public debates between "knowledge" and "skills" or "modes of historical thinking and reasoning" as potential goals for history education. Orientation in historical time does, of course, require a certain level of historical knowledge about events and circumstances that prevailed in different times; however it also involves the application of specific modes of thinking and reasoning. Proper history education will be well served by coherence between these two factors.

1

The Development of History Education

How has history teaching and the domain-specific educational science associated with it developed so far, and what role has awareness of historical time played in this? A brief historical overview may help in answering this question. The development of history teaching in three neighboring European countries is discussed: Germany, England and the Netherlands. This can help clarify to what extent certain developments affected the state of history education either in general or in a given country in particular.

The brief historical overview provided here is primarily intended to answer the question to what extent has awareness of historical time up to now played a role in history education. The overview is therefore not a detailed history of history education. Only the most important features of the developments in history education will be discussed.

The three countries were chosen because expertise in teaching and learning history based on empirical research is a well developed discipline in both Germany and England (in the English-speaking world in general, but England has been in the vanguard—see Seixas, 1994). However, both of these academic communities developed relatively independently from

Images of Time, pages 1–31

each other, with little knowledge of advances in the other community: very few English-language publications are cited in German works, and virtually no German-language publications are cited in English publications. The Netherlands has been influenced from both the English and the German side and may therefore be considered as a bridge between two worlds.

The development of history education is described in this chapter in three stages: first the developments up to the 1960s, subsequently the "crisis" in history teaching during the 1960s and the responses to that, and lastly the new societal and political interest in history education from the 1990s onwards. This division has been chosen because, from its origins up to approximately the 1960s, history education displayed certain characteristics that were subject to debate from the 1960s onwards. The 1990s displayed a reaction to the innovations of the 1960s and 1970s, which implied to a certain extent an inclination to return to the "good old ways" of history teaching. In this way, the response to the crisis in the sixties has determined current discussions on the essence and aims of history education. The role that awareness of historical time has played in history education so far is a point of interest throughout this chapter and will be discussed explicitly in the conclusion.

Developments Up to the 1960s

History as a general school topic made an early appearance in Germany: history was introduced as a compulsory subject in elementary schools in Prussia in 1763 (Ebeling, 1973). It was introduced into the elementary school curriculum in the Netherlands through the Elementary Education Act (lager-onderwijswet) of 1857 (Boekholt & de Booy, 1987). In England there has been resistance to state-managed education for a long time (Stephens, 1998), and schools were centered on churches and private institutions. History education blossomed within this context and was more or less viewed as an extension of religious and ideological instruction (Marsden, 1989). It was only after the introduction of the Elementary Education Act in 1870 that a national framework was created in England and Wales, followed in 1872 by a similar Act in Scotland (Stephens, 1998). History was not made a compulsory subject by these Acts. Prior to the passing of these acts, the influential Scottish academic Laurie had argued in his work On Primary Instruction in Relation to Education (1867) for a general adoption of history into the curriculum—a subject that to him would only be a waste of time if it was not used

> to bring to the schoolboy knowledge of wonderful deeds done in the discharge of patriotism and duty. History can bring him in bold relief the

> grand characters who through their heroic struggles for right and freedom have gained the privileges we now enjoy and by their great discoveries have made it possible for England to grow into a mighty empire. (as cited in Marsden, 1989, pp. 513–514)

History would gradually develop into a permanent feature within British elementary school education. The Board of Education, precursor to the subsequent Department of Education and Science, proposed in 1905 that "there are strong reasons why an important place should be given to history in the curriculum of every school" (as cited in Sylvester, 1994, p. 9). What were these "strong reasons"? What were the aims of history education during the hundred year period from approximately 1860 to 1960? A number of categories of reasons may be distinguished, which are discussed in order below.

Ethical Aims

In the first compulsory history lessons, ethical aims were considered to be of great general importance. For instance, it was said that the aim of introducing history in Prussian schools was "the understanding of things which belong to the beatitude of human society" (as cited in Ebeling, 1973, pp. 19–20, my transl.). The father of German history education, Kohlrausch (1780–1867), extolled the virtues of history education with arguments inspired by the Enlightenment. According to Kohlrausch, before the teacher sends his students off into the world, he has to teach them wise lessons, warnings and exhortations, wherever he can: "In the great domains of history he shows them, what they have to hold on to, what can invigorate them, what they have to look up to over and over again not to lose courage in the hustle of life" (as cited in Jeissmann, 1978, p. 66, my transl.). In England Newbery published his history book for his "young friends" in 1763 so they would become "wise and happy" through the examples of others, "which is the true end of history" (as cited in Reeves, 1980, p. 2). William Mavor wrote a collection of biographies for young readers in which they were encouraged to enter the "the temple of honour by the path of industry and perseverance" through learning by example from others (Reeves, 1980, pp. 2–3).

Considerations such as these were also to be found in the Netherlands. In Burggraaf's *Aanwijzing voor 't Onderrigt der Aardrijks en Geschiedkunde in de Lagere Scholen* (*Recommendations for Teaching Geography and History in Elementary Schools*) from 1824, he stated that the aim of history education for the lower classes was "to help them be prosperous, useful to others and in following God's commandments to be uncomplaining, satisfied, content and grateful" (as cited in Toebes, 1976, p. 207). The school principal Hemkes opined in his guide for new teachers from 1844 that history education was useful "whilst it serves to shape, glorify and refine the human spirit, enriches the soul with beautiful and useful thoughts, combats and roots out prejudices, destroys superstitions, cultivates patriotism and leads to the knowledge and worship of Our Lord" (as cited in Mulder, 1990, p. 216). It was noted in the memorandum to the Dutch 1878 Schools Act that for history in schools, "only that part of the nation's history providing the student with a comprehensive overview of the birth of the Netherlands State and teaching him the great deeds of our ancestors, whose actions under the leadership of [the Prince of] Orange established our independence, deserves further development" (as cited in De Jong, 1977, p. 10). Our ancestors were there to inspire the youth. Moral development was an important aspect of history education, as also emphasized by Kollewijn, author of history books for middle schools (*Hogere Burger School*) in 1889. Exclamations from students such as "that was plucky!" or "that was vicious!" were to be considered as evidence "that one had managed to stir the child's moral sense within his spirit" (as cited in Amsing, 2002, p. 279).

These goals of early history education may be summarized as the formation of morally upright individuals who would become useful members of society. This was to be achieved through learning about the exemplary and heroic behaviors of people from the past. At the end of the Second World War there was a return—with some emphasis—to this type of history education in the German Federal Republic. Once permission had been granted by the allied authorities in 1949 to restart history education, a flight away from political history to "*Kultur- und Menschenkunde*" (cultural and human studies) was observed (Toebes, 1981). This was characterized by ethical aspects in particular. Personal development and simple virtues such as obedience, selflessness, love, faithfulness, truth, level-headedness, and industry had to be worked on through following inspirational examples of people in the past (no longer heroes and statesmen, but artists, scientists, philosophers, and "ordinary people") (Toebes, 1981; Bauvois-Cauchepin, 2002).

Political and Civic Aims

Kollewijn did not only reserve his praise for the influence of morally upright examples in 1889. His description of the aims of history education also had another dimension:

> ... in order to be become a true human being, to observe one's social and civic duties, to live as prosperously as possible—something which all strive

for—one must to some extent have knowledge of our current, complex social and civic relationships. In order to be able to understand the present to a certain degree, one must know how it was created by the past, and there is only one subject from which we may learn this: history. (as cited in Amsing, 2002, pp. 277–278)

Increasingly greater numbers of the Dutch population were becoming full citizens through the expansion of the right to vote. History could teach them the intricacies of political life—a life in which they would have a future role to play. They would have to learn how to deal with their rights and obligations in a responsible way.

In the same year Kollewijn published his leaflet, the German *Kaiser* Wilhelm II in person detailed guidelines for history education in an *Allerhöchster Erlass* (supreme order). Schools had to make it clear to children that social-democratic learning was not only in conflict with the Divine Order and Christian ethics, but also unworkable in reality, and that it would have pernicious consequences. The power of the state alone could protect the freedoms and rights of the individual and his family. Youths had to learn how the Prussian kings had ensured them from the time of Frederick the Great's reformations to the present (Ebeling, 1973).

The obligations of the good citizen are also evident in the English sources referred to previously. These aims of history education may be summarized as follows: cultivating an understanding of the future role of students as citizens of the state through insights into its historical development and encouraging them to responsibly observe their obligations in this respect.

Nationalistic Aims

One specific political goal was to stir up nationalism. Although Dutch school principal Hemkes referred to nationalism as one of the explicit goals of history education, and although the preamble to the Netherlands 1857 Schools Act talks of "encouraging a warm sense of patriotism as part of the national curriculum" (as cited in Toebes, 1976, p. 207), nationalism developed less explicitly within the Netherlands during the nineteenth century than in the neighboring countries. In a methodology book for German education from 1889, the author Rusch saw the particular value of history in the shaping of patriotic convictions (Glöckel, 1973). His colleague Rude wrote in 1914: "In the first place, history education should strive for a real, decidedly German spirit. If this is not achieved, it has failed its most splendid goal. The schoolboy already should be proud to be a German" (as cited in Glöckel, 1973, p. 17). Faithfulness until death had to be demonstrated by means of good examples, and it had to be emphasized that many important inventions, such as the printing press and the telephone, were of German origin. During the Weimar Republic government policy aimed at moderating nationalism and paying more attention to the formation of a civic society in general. However, conservatively inclined teachers refused to heed the measures and continued to promulgate the old nationalistic line (Bauvois-Cauchepin, 2002). It is, of course, obvious that these aims were also pursued within national socialist Germany (Glöckel, 1973; Bauvois-Cauchepin, 2002).

Similar themes were also present in English history education. There was a virtually exclusive focus on British history at schools: "with the object of showing the emergence of Britain as a great power as well as the guardian of liberty," the "unique qualities of the British, stability at home and success abroad" was emphasized (Carpenter, as cited in Toebes, 1981, p. 162). Throughout the entire nineteenth century the aim of history education was to create "good citizens." This was rooted in an anti-papal Protestantism, which in the years prior to the First World War and under the influence of imperial developments and social Darwinism had begun to show signs of patriotism, worship of heroes and even racism (Phillips, 1998). The horrors of the First World War did create some reaction against excessive nationalism, yet Anglo centrism and patriotism continued to dominate the curriculum: "the notion still prevailed that patriotism and imperialism, so long as they were of British variety, were the keys to world understanding" (Marsden, 1989, p. 522). These traditional goals remained virtually unchanged until the nineteen sixties (Little, 1990).

The *schoolstrijd* (literally, "school battle," an ideological battle in the 19th century Netherlands between the providers of religious inspired private education and neutral public education) was influential in developing Protestant and Catholic forms of history education in the Netherlands, next to a "general," neutral variety. The difference in the amount of focus paid to the (Catholic) Middle Ages and the (Protestant dominated) seventeenth century is conspicuous in the schoolbooks of the various denominations (De Jong, 1977). The church in the Middle Ages, the Dutch Revolt, the role of William of Orange and the growth of the Republic under the leadership of the House of Orange provided so many points of contention between the Catholics and Protestants, that the (neutral) school principal previously cited despaired in his guide for new teachers from 1862 that religious organizations "only regard history from a Protestant, or Roman Catholic, Christian or Jewish perspective, and seem unable to present people simply as the human beings they are" (as cited in Mulder, 1990, p. 222). The socio-religious approach to national history would continue to remain the vanguard in the Netherlands up to the 1950s and 1960s. For instance, the foreword to a Protestant Christian history book from 1950 reads as follows:

> We hope that our simple work will contribute to a deeper understanding of national identity in the young and old. We believe that national identity may only have real value when it is grounded in the conviction that God's work is present and will continue to be present in the Netherlands until the end of time. (De Fouw, Van Bennekom, & Van de Bunt, 1950, p. 3)

Ideological approaches to history education under the influence of religious denominations were not a unique Dutch experience. For instance, an English methodology book from 1860 explains how working with contrasts may help clarify historical subject matter, for example by comparing the reign of Mary Queen of Scots with that of Victoria. The first was "despotic and cruel, bigoted and intolerant, miserable, a blind romanist, a friend of ignorance and superstition." In contrast, the second was "liberal and benevolent, pious and tolerant, cheerful and happy, an enlightened protestant, a promoter of education and religion" (as cited in Marsden, 1989, p. 512).

The aims in this category may be summarized as follows: creating a sense of love for and pride of one's own people, or one's own group within that population, and demonstrating by means of history that this nation or group of people have always served the right cause.

Alternative Aims

Attempts were made during this period in each of the three countries to counteract the mainstream of political-ideological and nationalistic approaches to history education. *Reformpädagogik* was developed in Germany during the Weimar Republic, which attempted to shape education to be more *"vom Kinde aus,"* or from the perspective of the child (Ebeling, 1973; Schuthof, 1983). For history education this meant propagating the use of stories and images. In England, Keatinge published his *Studies in the Teaching of History* in 1910. He wanted students to learn about research methods used by the historian and presented "problems" to students for them to solve (Aldrich, 2006). His example was followed by Happold in 1927 when he published his pamphlet *The Study of History in Schools as a Training in the Art of Thought.* According to Happold, history teaching needed to focus on

> the ability to collect, examine and correlate facts and to express the result in clear and vivid form, freedom from bias and irrational prejudices, the

> ability to think and argue logically and to form an independent judgement supported by the evidence which is available, and, at the same time, the realisation that every conclusion must be regarded as a working hypothesis to be modified or rejected in the light of fresh evidence. (as cited in Aldrich, 2006, p. 161)

In 1928 Happold published his *Approach to History* in which he set out his vision of history education: active students, projects, and problems. Other reformers attempted to focus attention on social history and diachronic themes such as housing, clothing or transport through the ages. From the fact that people deemed this type of history particularly suitable for the "less able," it is apparent that pedagogical considerations, such as "child-focus," played a role in this (Toebes, 1981, p. 163). The high-minded ideals of someone like Happold were certainly not mainstream during the 1930s. According to Worts, principal of the City of Leeds School, the school at which Happold had carried out his experiments could not be considered "as a normal standard for the average secondary school" (as cited in Aldrich, 2006, p. 163). Worts argued for "stimulating interest, adding color and infusing the sense of action and reality" (as cited in Aldrich, 2006, p. 163) in history teaching instead of critical studying sources.

Also in the Netherlands there were reformers in the first half of the twentieth century who argued for a new approach to history education, such as Te Lintum, De Vletter, and Albers (Toebes, 1976). These teachers argued that greater account needed to be taken of the stages of development of the students. Also, they wanted to break up the primacy of political history, and argued that more attention should be paid to social and economic history, daily life and ordinary people. According to Albers, attempts also needed to be made to develop students' capacity for independent critical thought. However, these attempts at modernization came up against a firmly ingrained tradition and often against evident resistance from teachers, particularly those teaching at denominational schools (Toebes, 1976). The call for more images in education, also based on pedagogical considerations, was, however, honored with the creation of historical images and associated stories for schools. These were mainly used in elementary education.[1]

The aims of the educational modernizers of the first half of the twentieth century, who were unable to exercise any significant influence on mainstream history education, may be summarized as follows: consideration should be paid to the potential and interests of the child and therefore there should be more history about daily life than about politics or wars. Some modernizers (e.g., Happold, Albers) also recommended presenting history as a subject that deals critically with information and aims at creating objective images of the past.

Conclusion

The development in history education up to the 1960s in the three countries demonstrates remarkable concurrences. History education was subject to ethical, political/civic and nationalistic agendas. Some modernizers attempted to present a different voice; however, this did not have a major impact on mainstream education.

The development of an awareness of historical time does not seem to have played a significant role during this first century of universally compulsive history education. The goals of history education were orientated towards the present. Examples of great deeds from the past had to be related to the present, with little attention paid to the distance in time and differences between the past and the contemporary age. History was utilized to present eternal values and truths rather than focusing on their variability and relativity. This was promoted through nationalistic interpretations of history. The existence of the present day nation as an "imagined community" was projected back into the (distant) past (Anderson, 1991). German children were taught to view Hermann's defeat of the Romans near the Teutoburg Forest of 9 AD as a German feat (Bauvois-Cauchepin, 2002); Dutch children were taught that the Batavian Uprising against the Romans of 69 AD was an early example of the Dutch sense of liberty (Teitler, 1998); and British children were taught that Boudicca's revolt against the Roman invaders of 60–61 AD was an example of heroic behavior by the British people (Williams, 2009). Rude's 1914 methodology book lists the names of heroes and statesmen who had to be studied. It included, in particular, Siegfried, Frederick the Great, Blücher, and Bismarck (Glöckel, 1973). Historical consciousness was evidently not an important aim in studying these heroes, if only because of the presence of the name of Siegfried on this list.

Anachronistic thought, rather than an awareness of historical time, was the order of the day. The difference in the focus on the Catholic Middle Ages and the Protestant 17th century in Catholic and Protestant history books in the Netherlands speaks volumes, demonstrating the rectitude of one's own socio-political group in the present, rather than casting light on the roles played by religions in the past. The demands of personal and political development are diametrically opposed to an objective and unattached view, which is the hallmark of a scientific approach to history (Grever, 2007a).

> It was the purpose of teaching history to produce subjects, later on loyal and responsible citizens, who would absorb—through the study of a highly selective system of historical knowledge—the ideology of history, the one and only history of the state. Thus history teaching ended up in an objective contradiction with historical scholarship, which boasted freedom to choose its objectives and objectivity as its regulatory ideas. (Bergmann, 1976, p. 3, my transl.)

But even the "historical scholarship" in the nineteenth century, and partly the twentieth century, has been significantly nationalistic as a rule (Blaas, 1988; Jonker, 2001a; Geary, 2002). So the objective contrast between history education and historical sciences, as suggested here by Bergmann (1976), could be of little import for a considerable period of time.

The Crisis of the 1960s and the Responses to It

A study among students in secondary education commissioned by the British government in 1966 showed that they placed history at the bottom of the list of subjects that were "useful and interesting" and almost at the top of the list of subjects that were "useless and boring" (Price, 1968, p. 343). Many British history teachers were in a state of "perplexity and pessimism" (Price, 1968, p. 342) about the situation of their subject. British students did not hide their antagonism towards it: "With change proceeding in a nightmare progression, of what use to me is the experience of people born before the Second World War?" "How can anything before the Space Age teach me to cope with it?" (Reeves, 1980, p. 3). A similar picture was painted by a German study of students conducted around 1970: "the old times do not interest me any more. We are in a different time now" (Wiesemüller, 1972, p. 32, my transl.). Some German educational publications from this period point in the same direction (Süssmuth, 1972). The amount of history in the curricula of the diverse German states was reduced in favor of integrated social sciences; in some cases, history was completely absorbed by social sciences (Kocka, 1977). In the Netherlands, the number of history lessons in the foundation years of secondary schools was significantly decreased in the context of the Secondary Education Act of 1963; in the curriculum of the higher grades of secondary education, history was reduced to an optional subject. According to the historian Pieter Geyl, who wrote a disquieting article in the periodical *Vrij Nederland*, the Inspector-General for Education, Goote, had informed him that "little value could be attached to history as a school subject in a dynamic time such as ours" (as cited in Toebes, 1981, p. 251). It is clear that history, which for so long had been able to boast of highly relevant political and ideological goals set by politics and society, was now in crisis.

The crisis of history in the schools reflected a general decline in the esteem of the subject. In Germany, there was reference to a *"Verlust der Geschichte"* (loss of history), *"Geschichtsmüdigkeit"* (tiredness with history), or even *"Geschichtslosigkeit"* (being without history) (Toebes, 1981, p. 62). In England the historian Plumb published his book *The Death of the Past* in which he argued for a new role for history: "the past used to dictate what a man should do or believe; this history cannot do..." (Plumb, 1969, p. 15). Rather than making people aware of the ties binding them to their ancestors, modern historical science was showing them what was alien and different in the past. Lessons from history therefore became less clear cut than they used to be: "Bygone feats would soon cease to justify tyrannous authority, to affirm self-serving manifest destinies, to incite citizens to voice malign vainglory" (Lowenthal, 1998, p. 90). The social sciences offered different, perhaps more valuable, opportunities to explain the world. In addition to this, Germany in particular had to deal with its most recent history, a difficult topic to handle, especially for the generation of students being confronted with the guilt of events to which they had not been party. In the United Kingdom the dissolution of the British Empire played a role in the increasingly more awkward message of "Britain's greatness" (Ward, 2004).

Responses to the Crisis

History Teaching is About Historical Awareness

The attacks on history as a school subject resulted in a re-evaluation of its aims and content. At the German *Historikertag* (Historians' Conference) of 1970, which was devoted to the theme *Wozu noch Geschichte?* (Why History Yet?), the historian Kocka (1977) presented seven functions that history could and even perhaps should fulfill in a modern liberal-democratic society:

1. Understanding of present-day phenomena by means of studying their origins and development in the past.
2. The awareness that there have been many more types of human existence in the past than just those of the present; this means that someone's insights may be expanded by studying not only analogies between the present and the past or within the past, but also examples from the past that are diametrically opposed to situations in the present.
3. A critical appraisal of the role played by traditions and myths in society.

4. An understanding of the changeability and variability of human societies (including those of the present), as well as an understanding of the power of status quo, which counteracts changes being carried through easily.
5. The understanding of human nature provided by the study of human societies other than one's own, such as societies that have existed in the past.
6. Insight in the complex relationships between dissimilar factors in a historical process, where unique circumstances sometimes play a role; insight into multi-causality and contingency in history.
7. An understanding of the totality of human existence where other subjects often focus on one specific aspect (Kocka, 1977).

History, according to Kocka (1977), suffers under dictatorships or other forms of authoritarian regimes. In contrast, its contribution to a liberal-democratic society is essential; it is a contribution that history can only give up "at the price of its own self-given task" (p. 22). History and liberal society are mutually dependent upon each other. Kocka attempted to derive the functions that history should have within a liberal, democratic society from history as it develops within such a society. This was a fundamentally different approach to the "imposing" of aims by governments as had been the norm in traditional history education—aims which, as has been made clear in the previous section, were often in conflict with the objectives of a critical and scientific approach to history and were often at odds with the development of an awareness of historical time.

Why and how Kocka's seven functions are essential to a free and democratic society can be imagined without too much effort as we reflect on them. Historical thought not only demonstrates that conditions used to be different; it also demonstrates that developments in the past might have taken a different course. This allows us to take a detached view of matters as they currently are. Therefore, historical thought opens the way to critical thought and consideration of alternatives. History demonstrates that the things people believe in and human values are changeable and relative, that people change their opinions. This insight may help promote open debates in a democratic society. Kocka deliberately renounced mentioning the shaping of identity as a function of history, precisely because the critical views yielded by a scholarly history are more likely to produce distance from than a feeling of alliance with ancestors from the past.

Thus, the crisis of the 1960s resulted, in Germany, in historical thinking being taken seriously as an aim of history education. Historical awareness of time played an important role in this, as is apparent from Kocka's seven functions, which to a significant extent result from insight in time differences. According to Rüsen (1994), learning history comes down to "a process of human consciousness in which certain experiences of time are appropriated interpretatively while at the same time the competence of time interpretation originates and is being developed further" (p. 64, my transl.). These operations of human consciousness involve dealing with time in a way that is characteristic of history. The very fact that this type of thought occurred in Germany is probably related with German historicism and the work of thinkers such as Husserl and Heidegger, who had been extensively occupied with the temporal aspects of human existence and human culture (see Chapter 2). The awareness of the existence in time, the origins and future of man's creations, and the variability of these creations are, by Rüsen's definition, important components of historical awareness. These are topics to which Kocka also referred in his analysis.

History education in which categories such as those highlighted by Kocka were to play a guiding role tended to concentrate on the choice of specific subject matter and, in particular, on the way in which this historical content was presented (Toebes, 1989; Wilschut, van Straaten, & van Riessen, 2004). For instance, certain topics could serve to demonstrate changeability or the power of status quo, others to create an insight into complicated causal relationships or the unpredictability of the course of history, and others again could be used to draw comparisons, analogies and differences.

The German schoolbook series *Fragen an die Geschichte* (Questions to History), published from 1974 onwards, is an example of teaching material created according to these principles (Schmid, 1972, 1974). The title shows how questions were used to provide a criterion for the selection of subject matter in such a way that history could play its essential role in a liberal society. To elucidate how his educational principles differed from previous ones, the author Schmid presented pairs of "traditional" and "innovative" teaching objectives for history education. An example of a traditional teaching objective was "Understanding of the religious and political factors that determined a crusade"; the innovative objective running parallel to this read, "Can a Christian justify a military mission on the basis of his convictions and beliefs?" Another pair of objectives mentioned was: "Name the reasons for the discovery of America" and "What may motivate someone to risk his life to discover new worlds?" (Schmid, 1972, p. 233). The selection of subject matter would no longer be focused on the transfer of cultural heritage, but would come about via questions targeting problems relevant to people of the present. In this way, in spite of the promoting of a historical awareness of time that could be the result of this kind of teaching historical thinking and reasoning, history education yet again ran the risk of becoming a too much present-oriented trade.

History Teaching is About Studying Evidence

The English response to the crisis of the 1960s was deeply influenced by two American educational publications: the *Taxonomy of Educational Objectives* by Bloom & Krathwohl (1956) and Bruner's visionary work (Booth, 1994). In Bloom's taxonomy there were five higher categories of educational objectives in addition to the basic one aiming at knowledge: understanding, application, analysis, synthesis, and evaluation. Bruner (1960) believed that any topic could be taught at an acceptable level to any group of students provided that students were introduced to the basic concepts and principles underlying a discipline at an early stage: "the basic themes that give form to life and literature are as simple as they are powerful" (pp. 12–13). The basic concepts and principles had to be re-presented continuously at an increasingly higher level via a "spiral curriculum" (pp. 52–54).

Bloom and Bruner generated questions to which history education did not have an immediate response. If history education was not just about knowledge, but also aimed at those five other categories of objectives, what type of history lessons would that produce? How could a higher level be reached in history other than by memorizing more knowledge, increasingly more complicated knowledge? What would a "spiral curriculum" look like in the context of history? What were the basic concepts and principles that needed to be presented at an early stage and then continuously re-presented at increasingly higher levels? Could something like this be achieved by a chronological curriculum, which dealt with a sequence of continuously differing historical events and situations?

In the United States these questions were addressed via the renewal of the "social studies" curriculum, of which history was only a part and often a subordinate part (Seixas, 1994). The "new social studies" could no longer be involved in the memorizing of knowledge about social reality—knowledge which would quickly become outdated. True to Bruner's assumption that any subject could be taught at any level, as long as the basic principles were studied at an early stage, there was a search for footholds in academic disciplines and their associated research methods. The basic principles had to be discovered independently by students by working with data and texts instead of with books containing all the answers. The process of discovery needed to be modeled towards "the mode of inquiry of history and the social sciences, to the manner in which practicing social scientists develop and validate hypotheses" (Fenton & Good, 1965, p. 207). The skills acquired during this process could continue to be used for different topics after formal schooling had been completed.

Coltham and Fines (1971) developed, within this spirit, a set of objectives for history education in England. The influence of Bruner and Bloom and thoughts about the "mode of inquiry" from the new social studies originating in the United States are clearly evident in this document. The chapter on the *Nature of the Discipline* only refers to objectives concerning the processing of evidential material that may be found in historical sources. This shows that history, just like the social sciences in general, was being characterized as a subject in which critical processing of information played a central role. In the chapter *Skills and Abilities* the categories from Bloom's taxonomy were applied only to dealing with historical evidence. Little attention was paid to historical awareness and historical thought as described by German researchers. Questions such as "How do I ask the right question or formulate a good hypothesis?", "How do I assess the reliability and representativeness of my evidence?", "How do I deal with bias in my sources?", "What can I accept as fact?" and "Is it possible to be objective?" were central to this approach (Fenton, 1967, pp. 15–16). Yet these are questions that could apply to any social science, not just to history alone. They have little or no concern with awareness of historical time; they do, however, introduce a certain level of distrust in traditional, authoritative information as had been presented as standard in history lessons up to that point without discussion. The impetus to ask questions was the new political message presented to students during the 1960s.

An example of this new approach to history education was the English *Schools Council History Project* (SCHP), which was aimed at the first three years of secondary education. This project, which did not involve all English schools, attempted to tailor history more towards the "needs of students" in order to enable them to better understand the world around them (Sylvester, 1994; Schools Council History 13–16 Project, 1976a). The traditional chronologically ordered curriculum focusing on British national history was rejected by the SCHP. History was not to be considered as a chronologically ordered entity of knowledge, but as an endless quantity of poorly structured details about what people did in the past. History was a mode of inquiry aimed at processing these data (Schools Council History 13–16 Project, 1976a). The curriculum therefore needed to be based on subject-specific research methods, or historical sources being studied as evidence on which facts were based. By placing concepts such as "evidence," "question," "fact," and "objectivity" and the associated cognitive skills at the center, the higher levels of understanding as proposed in Bloom's taxonomy could be reached. Education could be ordered into a spiral curriculum by continuously returning to these skills and basic concepts at increasingly higher levels. The historical content used in this process played a secondary role. A significant amount of the *Schools Council History Project* subject matter therefore appeared to be quite traditional. For instance: contrasting evidence had to be studied or analyses of causes and consequences had to be made concerning good old topics like Elizabeth I or Henry VIII and his wives (Sylvester, 1994).

It is apparent from the teaching series *What is History?* (Schools Council History 13–16 Project, 1976a), which was used to introduce history education along the plans set out by the SCHP, that the study of evidence was the dominant theme, and the focus on awareness of historical time subordinate. Three out of the five units from this series were devoted to studying evidence (numbers 2–4) and one to asking questions (number 5). The first, about "people from the past" served to explain how years and timelines worked. In contrast to the units about working with evidence—all of which consisted of a brochure of 25 to 45 pages—the unit about time and timelines consisted of a single leaflet (Schools Council History 13-16 Project, 1976b). It is clear that dealing with time was not considered a topic that required a great deal of attention.

The influence of Bloom and Bruner was not explicitly noticeable in Netherlands in the first Objectives Report issued by the Dutch Association of History Teachers (VGN, 1967). The report did link up with the general trends in educational modernization that were prevalent in the 1960s, such as promoting students' independent activities in terms of thinking, feeling and handling information. The "old school" placed too much emphasis on studying knowledge and the transfer of information. Instead, students had to develop skills to acquire this knowledge and information themselves (VGN, 1967). This involved an orientation in the world of the present, rather than the world of the past. The fact that a lot of history education did not deal with developments after 1918 was not particularly helpful in this respect (Van der Werff, 1973). Therefore a new examination program was introduced in 1965 that turned things upside down: only the "last fifty years" were to be examined (in practice this meant historical developments since 1917), with the choice of themes from previous centuries left to the teacher (Fontaine, 1982).

The American influence did become more noticeable in the Netherlands during the 1970s. Dalhuisen, an expert in teaching and learning history from Leyden University, published a leaflet promoting the "mode of inquiry," which was virtually a literal translation of what the American Fenton had argued within the context of the "new social studies" (Dalhuisen & Korevaar, 1971). Dalhuisen was the editor of both editions of the handbook for teacher education *Geschiedenis op school* (History at School), which appeared in 1976 and 1982-1983 respectively (Dalhuisen, Geurts, & Toebes, 1976; Dalhuisen, Toebes, & Verhagen, 1982). A special chapter written by Dalhuisen was devoted to the "method of inquiry" in the 1976 edition, while the chapter on "objectives"—also written by Dalhuisen—clearly demonstrated influences of Bruner and Bloom, as well as the new social studies from the United States.[2] The study of evidence to be performed autonomously by students was central to this.

Gieles was another Dutch expert in teaching and learning history to develop Bloom's taxonomy. The study of historical evidence, however, did not feature in his approach (Gieles, 1973). Gieles was of the opinion that the activities of professional historians had no role to play at school:

> Little attention is paid in history lessons to the important task of an historian, that is of collecting facts and assessing their veracity... That there is so little attention to this and other tasks of historical science is associated with the fact that in the Netherlands history teaching is not intended to introduce students to professional studies, but that the results of these are used for educational aims. And that, we believe, is correct. (Gieles & Greep, 1979, p. 30)

Thus, Gieles aligned his argument to the classical aims of history teaching that had primarily focused on knowledge and general education.

Developments in the "New History Teaching" in the 1980s and 1990s

After the initial concentration on studying historical evidence, purer historical concepts and categories were introduced into educational research in England during the 1980s and 1990s. A pure historical concept, for instance, was the idea that students had to transfer themselves to different points in time in order to understand and judge the past from a viewpoint in the past. It was in a sense unfortunate that the term "empathy" was chosen for this in the English literature, creating confusion with "sympathy," with the intuitive feeling for people from the past rather than the intended rational cognitive proceedings involved in historical understanding. This led to a great deal of misunderstanding and criticism (Lee & Ashby, 2001). What was at stake was the development of the students' ability to view things from different perspectives, namely perspectives from the past. Awareness of historical time played a role in this.

This aspect of history teaching first caught the attention of Ashby and Lee in 1987. In the 1990s they conducted their research in the context of an ambitious long-term project *Concepts of History and Teaching Approaches*, or CHATA (Lee, Ashby, & Dickinson, 1993; Lee, Dickinson & Ashby, 1997; Lee & Ashby, 2000, 2001). Assessing historical evidence and formulating interpretations based on this continued to play a role in this project, in addition to issues such as insights into historical changes and explaining causes and consequences, including the aspect of transferring oneself in time to the context of events in the past. However, the research into the ways in which students dealt with evidence from the past not so much focused on the question of how sources from the past needed to be located within the context of their time; rather Lee and Ashby were interested in the question of how—according to students—a reliable image of the past could be formed "if nobody is around who had lived through it." The significance of processing information, as professed in the social sciences, remained at the forefront (Lee & Ashby, 2000, p. 212).

Not only in educational research, but also in educational practice, the study of historical evidence continued to play an important role in England, partly due to the fact that it was introduced as a general examination requirement for *General Certificate of Secondary Education*, or GCSE (Department of Education and Science, 1985). An impression of the emphases in the GCSE program can be derived from the periodical *Hindsight*, which was developed for students preparing for the GCSE examinations. It included examples, tips, tricks and additional teaching material. The regular feature *Improve your grade*, which aimed at training for better examination results, included a majority of examples of evidence studies. The much smaller number of examples dealing with concepts such as causality, continuity, change and interpretations also mostly consisted of collections of fragments of evidence based on which students had to form their own image of the past (*Hindsight*, vol. 1–8, 1990–1998). According to Stow and Haydn (2000), the GCSE program resulted into the cohesive story reflecting the past taking a backseat: "It would be possible to gain maximum marks in a source-work question on the Second World War, without necessarily knowing how the events of the war unfolded, who was on whose side, or who won" (p. 84).

In the Netherlands the study of historical evidence was given a permanent berth in the curriculum next to a few purer historical modes of reasoning. In 1986 a report was published by the Commission for Revision of Final History Exams (HEG), in which Dalhuisen played an important role (HEG, 1986). In contrast to the English examinations, the centrally organized written history examination, introduced in 1981 in the Netherlands, initially did not test any of the higher categories of Bloom's taxonomy, as the committee introducing these examinations was of the opinion that with a subject like history one "meets limits" here in this respect (CEGES, 1979, p. 45). This was not the position adopted by the HEG. Dalhuisen had developed his "mode of inquiry" into a program that he referred to as "structural concepts and skills" (Dalhuisen & Van der Dussen, 1975), which was included in its entirety in the HEG proposals. As a result the concepts and skills were introduced into the examination program in 1993 (WIEG, 1993). The ability to deal with historical evidence played an important role in this.

The English "empathy" was reflected in the Dutch categories *inleving* and *standplaatsgebondenheid* ("imagining oneself in" and "location-boundness," or more or less: perspective taking); the latter had been developed within the social sciences, with the foundations provided in the *interbellum* by the epistemic sociologist Karl Mannheim and his thesis of *Standortgebundenheit* of knowledge (Frisby, 1992). This thesis, in brief, maintains that all knowledge is bound to an origin, to the background of those introducing the knowledge and, as such, should not be considered as being independent. A description of the concept intended for the new history exam program according to the HEG proposals ran as follows: "Every human being has his/her own background. This is influenced by various factors, such as the time in which someone is living, the experiences in someone's life, the place where someone lives, the country, the gender and social status someone belongs to, someone's religion and political convictions" (WIEG, 1993, p. 16). Time dependency is referred to in the same breath as the general sociological dependency of humans to their social circumstances.[3] The category of *inleving*, like the English concept of *empathy*, was conceived of in a true historical way, but the *standplaatsgebondenheid* (perspective taking) remained a general social scientific category.

In the meantime, Toebes was advocating a modernization of Dutch history education inspired by German expertise in teaching and learning history. He launched a series of schoolbooks under the title *Vragen aan de geschiedenis* (Questions to History, 1983–1996). The fact that this title reflected its German counterpart published in 1974 was not a coincidence. In the introduction to the series, Toebes explained that the selection of subject matter, in particular those topics that could shed light on the influence of aspects of historical awareness as described by Kocka, was essential to good history education (Toebes, 1983). Toebes' educational approach, however, was not influential in the development of the final examination programs and attainment targets in the Dutch history curriculum (SPTF, 1995; Sniekers, 1993; WIEG, 1993). The trend established by Dalhuisen had set root and the predominant themes were primarily a command of the skills required for studying historical evidence and some historical thinking and reasoning; the historical content to which these had to be applied was considered to be of minor importance.[4] The impression that historical content played a secondary role was emphasized by the application of varying examination themes for the final examinations each year; since the renewal of the exam program introduced by the HEG, these no longer concentrated solely on history after 1917, but also included various themes from other periods. The selection of examination material thus seemed to have a certain element of arbitrariness, even though the committee reviewing the examination programs in the context of the introduction of a renewed upper stage of secondary education (grades 10–12) in 1998 attempted to introduce some streamlining in the selection of themes by introducing learning domains from which themes had to be chosen. The choice of these domains had been suggested in particular by the development of new sub disciplines in academic history (SPTF, 1995).

Thus, by the 1990s, history teaching in England and the Netherlands had been fundamentally revised as a result of developments sparked off by the crisis during the 1960s. The chronologically ordered political overview was replaced by a much more loosely ordered set of themes. Research skills derived from the social sciences had been added as an important component to the curriculum. In addition, some attention was also paid to reasoning skills based on an awareness of historical time. However, the latter was not to receive a guiding role—as advocated in Germany and by Toebes in the Netherlands—in the history curriculum in England and the Netherlands.

Conclusion

During the 1960s traditional history education was viewed as no longer being acceptable and sensible in each of the three countries. Traditional historiography had supported nationalistic tendencies and created a basis for moral and political education at schools. However, modern historical science was no longer prepared to accept this, as argued by Plumb (1969) in his *Death of the Past*. As a result, politics and society more or less turned their backs on history education. Social sciences, in which history could be subsumed, were believed to be able to provide a more useful curriculum. This led to two types of response among historians and history teachers. One was concerned with the role that historical awareness could and should fulfill in modern society. The other was an attempt to prove that history could be as useful to society as the social sciences through the critical study of sources and documents.

The first response was found in Germany in particular. In the search for the roles that historical awareness played in modern society it became apparent that thinking about historical time played an important role in this process. The particular way of thinking about reality that is referred to as "historical" led to topics being selected for a curriculum that were most suitable for promoting historical awareness. This had to involve subject matter that would be able to answer questions—from a historical perspective—from modern man by promoting an awareness of the historicity of human existence. However, as there was an explicit focus on present day questions and problems relevant to contemporary humans, this could have the effect of the awareness of the difference between the past and present, the awareness of an historical passage of time, becoming more relative.

The second response was dominant in England and the Netherlands, with a crucial influence from the United States. The modernization of "social studies" in that country had led to a focus on research skills. Although this approach had made history "useful" to modern society, it did not make clear why history should receive preferential treatment ahead of (other) social sciences. In contrast to this, the German approach was based on the roles that only history could fulfill optimally.

During the 1980s and 1990s, there arose within English expertise in teaching and learning history a focus on other aspects of historical thought next to studying historical evidence. Attention was paid to the diverse perspectives of different times that play an important role in explaining the actions of people and events from the past. This led to English expertise in teaching and learning history becoming more focused on awareness of historical time, although the critical use of evidence continued to be in the foreground.

The new history curricula led to the good old chronologically ordered (national) political histories disappearing from the stage. The historical contents of examination programs in England and the Netherlands became relatively arbitrary. The guidelines for the GCSE, as well as the new Dutch examination programs, allowed for a wide choice of historical themes. Students were starting to leave school without the well known package of historical knowledge with which the older generations had been familiar.

New Socio-Political Interest in History Education from the 1990s Onwards

During the course of the 1990s and the first decade of the twenty-first century, a revived interest in history could be observed (Frijhoff, 2010; Grütter, 1994; Jonker, 2006). This was linked to developments that had made the nation state ever less visible as the obvious context in which people were living. National boundaries were increasingly being transcended through the information revolution brought about by the Internet and other technological innovations, in combination with increasing levels of internationalization of economic developments. Migration on a mass scale accentuated the development toward multi-ethnic, multicultural societies. The political integration of the European Union intensified. The attacks of September 11, 2001 in New York were a symptom of growing tensions between the Islamic and Western Worlds. German unification in 1990 and the end of the Cold War demanded new historical perspectives. All these changes coupled with a kind of identity crisis among citizens of the Western World.

In the meantime, developments in academic history had led to a greater distance between history and the national context within which the study had always tended to figure. More important yet, perhaps, was the so-called "linguistic turn" in post-modern philosophy of history, which placed an emphasis on the fact that history only exists in the form of a number of representations in words: the only meaning that history can have is the meaning given to it by "narratives" (Ankersmit, 1983; White, 1973, 1987). This relativistic trend linked badly to the needs of those in search of the certainty of a clear, national identity in the face of a rapidly changing world.

A growing number of politicians were of the opinion that history could create a binding factor in an increasingly diversifying society in which national citizenship, national loyalty and integration were becoming increasingly harder to create. According to these politicians, national history would provide immigrants with a good introduction to the culture into which they hoped to integrate (Grever, 2007b). This advocacy of "nation building" seemed to bear some resemblance to similar attempts in the nineteenth century. There is some logic in the fact that similar tools were being re-utilized. Attempts were made to return history education to its traditional political and ideologically inspired role from before the 1960s. Additionally, voices were heard from those who believed that history education needed to link up in a different way to the new realities: global history, rather than national history and a multicultural perspective rather than national cultures (Phillips, 1998).

Return to a More National History in England and the Netherlands

The introduction in England of a National Curriculum, an initiative taken by governments under the leadership of Margaret Thatcher in the years 1979–1990, led to fierce debates about the "attainment targets" within it to be set for history as a school subject. The course of events in the 1980s and 1990s has been described in detail by Phillips (1998) in his *History Teaching, Nationhood and the State.* An interesting aspect within this is the role of politics as viewed by the participants in the debate. Various commentators attempted to belittle the "new history teaching" such as had developed during the 1970s and 1980s and to accuse it of political bias. Phillips (1998) names four of them: Beattie, Partington, Kedourie, and Deuchar. Partington contrasted the "mild socialist consensus" about history teaching in the 1950s and 1960s with the—in his eyes—highly politicized curriculum of the 1970s and 1980s: history had been "hijacked" by neo-Marxists. According to Beattie the new history teaching "was not really history," but rather "a watered down contemporary history" or "current affairs"; he also believed it had become politicized. Kedourie argued that history was "distorted," while Duchar denounced the "willfully perverse dismissal of historical content and facts" inspired by "a fundamental loss of faith in our civilization." According to Partington, a campaign was being waged which set out to preach "contempt and hatred for the central political and cultural traditions both of Britain and the western civilization of which it is part" (as cited by Phillips, 1998, pp. 32–33).

These arguments about politicization are diametrically opposed to the idea that it was rather the traditional history education that had been inspired by political ideology up to the 1960s, and that the "new history education" in the 1970s had been less influenced by ideological factors or influenced by them in a fundamentally different way. It is apparent, then, that the national political character of history education was experienced as being so self-evident, that the absence of it could be viewed as "politically motivated." This is not surprising, as history had become included in the curriculum with the aim of national education in the first place; it was its natural aim that it had always enjoyed. Direct political interference had not even been necessary to achieve this in England: "It was not as though central government felt obliged to promote, through the history classroom, a sense of citizenship rooted in English Protestantism. Essentially, history teachers in England did it for them" (Phillips, 1998, p. 14).

The harsh words that had helped critics in England to pour scorn on history education in the 1970s and beyond are understandable in this respect, that the debate was not carried within the professional arena, but in the public domain. Each phase in the development of the National Curriculum was accompanied by heavy-handed commentaries in the press, including occasionally screaming headlines (Jones, 2000; Little, 1990). An example of this type of commentary is the following:

> They have drawn up a history curriculum for schools in which crucial events in British history are—disgracefully—treated as optional extras. Instead, young people would be exposed to a 'correct' approach with an emphasis on gender and ethnic perspectives. In this fashionable lunacy it would be possible for a student to grow up knowing next to nothing about the Gunpowder Plot, Trafalgar, Waterloo or Winston Churchill. And everything about the experiences of black peoples in the Americas or the lifestyle of the Ancient Egyptians. (as cited in Crawford, 1995, p. 442)

The fact that the debate was carried out in the public domain also demonstrates how highly politicized it had become. This is also evident from the continuing interference by politicians in the work carried out on the National Curriculum (Crawford, 1995).

The debate in the Netherlands began at a later stage than in England. The first generations of attainment targets for grades seven through nine and the final examination programs for secondary education, which can be seen as counterparts to the English *National Curriculum,* were formulated without any social or political interference. However, in 1997 a Committee on History Education was installed by the Secretary of State for Education. Its task was to investigate which expectations of history education existed in society, including higher education. The committee came to the conclusion that: "an estrangement seems to have arisen between society's views on history and what is currently believed to be of importance in history education in the general sense, and has been translated into teaching objectives and introduced into teaching methods" (CG, 1998, p. 12). Society's expectations were heading in the direction of a return to traditional history education: more factual knowledge, broader chronological overviews. Voices calling for a more nationalistic approach to history had yet to be heard, although the word "canon" was introduced for the first time (CG, 1998).

A subsequent committee was instructed to develop the advice provided by the Committee on History Education into concrete proposals for attainment targets and final exam programs. This was the first time in the Netherlands that an attempt was made to create a single coherent program for history education from elementary education up to and including the higher grades of secondary schooling. The general expectation was that this committee—the Committee on Historical and Social Education (CHMV)—would produce a list of names, years and facts to be memorized.[5] However, this was not to be the case. Following the example of the German expertise in teaching and learning history from the 1970s and beyond, the committee proposed the development of historical awareness as the principal aim of history education. According to the CHMV, insights into chronology and characteristic features of eras would be useful in this respect (CHMV, 2001). The advice included the proposal to help develop students' chronological understanding via a curriculum of ten "easily remembered" eras, bearing "simple, associative names" (CHMV, 2001, p. 20). The committee was of the opinion that the choice of particular historical persons or events would be arbitrary and that a nationally prescribed program had to include only the general characteristics of eras to enable orientation in time (CHMV, 2001). The selection of examples of particular historical persons and events to be utilized for clarifying the characteristic features of eras should be left to schools, teachers and students. The chronological framework was developed to enable historical awareness and historical thinking, it was not intended to be utilized for the memorization of facts about (national) history. Therefore there was no reference to names, dates or significant events or heroes from national history.

This was not quite what an unsettled public opinion had been waiting for, nor was it something the Secretary of State had intended by creating both committees. When the CHMV report was released to the press, it was broadly misinterpreted as a proposal to re-introduce extensive compulsory factual knowledge into the curriculum.[6] The public interest in Dutch national history was on the increase, for instance in the form of the *National History Quiz*, which ran on public television from 2001, and the poll for the *Greatest Dutchman of All Time* (2004), following in the footsteps of its counterpart directed at the *Greatest Briton* (2002) on the other side of the North Sea, and at a later stage popular TV series such as *Holland's Past* and *Holland in twelve murders* (Vos, 2009; Van Vree, 2009). The call for identification with a "national heritage"—which immigrants also needed to adopt in order to become better integrated into Dutch society—was becoming increasingly louder. When the Education Council advised that a national "canon" could potentially contribute to a stronger Dutch cultural identity (Onderwijsraad, 2005) the Minister for Education, Culture and Science almost immediately created a Committee for the Development of the Netherlands Canon. This committee finally published the long awaited list: fifty individuals and events which "every Dutchman should know" (Van Oostrom, 2006).

The Netherlands Council of State advised against the compulsory introduction of the Canon, as this would conflict with freedom of education and was diametrically opposite to existing attainment targets and exam programs. The Council suggested to the government that the Canon should only be included as "a source of inspiration" in the preamble to the attainment targets (Raad van State, 2009). However, when the Minister for Education put forward this proposal to the Dutch Lower House, a motion was passed to replace the phrase "source of inspiration" with "point of departure for illustration" (Handelingen, 2009). An all too excessive level of disengagement with the national element in history education would, after all, not be in line with the ideological goals of education that by now had been adopted again by virtually all politicians.

Persistence of Historical Awareness in Germany

In Germany so far there has been no noticeable political pressure to create a more nationally oriented history curriculum. Nevertheless history education in Germany has also been confronted with new challenges. In September 1999 the former federal presidents Herzog, Von Weizsäcker and Scheel issued an appeal for a greater sense of community spirit. This appealed in particular to the former history teacher Henning von Burgsdorff, who felt that proper history teaching should make a contribution. The *Henning von Burgsdorff-Stiftung* (Association) was set up in 2000 with the aim of promoting historical awareness among the younger generation, particularly with a view to better citizenship (Netzel, n.d.). The organization supported the publication of two books about Germany's recent past that were "comprehensible to young readers." These were subtitled "how we became what we are" (Hachtmann, Rohlfes, & Ullrich, 2006; Alter, Rohlfes, Wolfrum, & Wolle, 2006). Another recent publication bears the same subtitle (Wiegrefe & Pieper, 2007). The books attempted to say something about German identity, but Wiegrefe and Pieper (2007) also emphasized that German history was "complicated" and that the nation state had not been an obvious outcome of historical developments:

> More than 50 years after the end of the war, history recurs into the perspective of Germans in a novel way. Its variable process of development with its errors and triumphs shows: the nation state is not the natural result of historical circumstances, but an invention. (p. 14, my transl.)

Education programs in Germany are set for each of the federal states separately, as education is not under federal control. Nevertheless, national standards for education are discussed in consultation between the ministers for education for the various states. The *Verband Geschichtslehrer Deutschlands* (VGD, Association of German History Teachers, 2006) developed a proposal for educational objectives for history in the secondary education curriculum of all states of Germany. Nine contributions of history to general education are highlighted in the introduction:

1. Cultural memory: history represents what we remember as a community, and what is important for us to be able to understand events of the present.
2. Clarification of the present: modern conditions have not arisen by accident or by fashion; rather they have grown through historical developments.
3. Orientation on the future: future developments will be the result of (occasionally long-term) historical developments.
4. Educational value of earlier periods: people have existed under circumstances that are very different to those facing us today; it is instructive to learn how they dealt with these conditions.
5. The value of experiencing something different: history confronts issues that at first sight appear to be incomprehensible. It is useful to attempt to understand these issues; this will help promote tolerance, among other things.
6. Multi-perspectivity: history is a discipline that deals with information from numerous sources and perspectives.
7. A profound liberal education: there are so many historical views, speeches and other matters present in our cultural world that we need some knowledge of them.
8. Interdisciplinary approach: history views humankind in its totality, not just a single aspect of being human, as is the case for the majority of other subjects.
9. Identity: a person's identity is shaped by his/her personal history. The same applies to community identities—communities to which an individual belongs, and to which his/her personal identity is bound—which are shaped by their history (VGD, 2006, pp. 9–12).

The influence of functions of history as described by Kocka (1977) in the 1970s and subsequent German publications concerning historical awareness is still clearly evident in this list. The topic of identity is a new addition, but it is treated discreetly. National or German identity or a national German heritage are not referred to by name. Nor do the learning standards—which, in terms of content, take a thorough and traditional approach—reveal any sign of a return to a national German history. The only specific focus on Germany is present in the subject matter on the 19th and 20th centuries: the German Empire since 1871, the Weimar Republic, National Socialist Germany and developments in Germany since 1945. These topics are placed within the context of general European or world history: imperialism, the world wars, the Cold War and European unification.

Conclusion

In the 1990s national history with its traditional, political/ideological objectives and aims of creating a civic society returned to the agendas of history education in England and the Netherlands. A confrontation arose with the new history education that had been developed during the 1960s and 1970s, which aimed partly at dealing critically with historical evidence and partly at historical awareness and historical reasoning. This confrontation did not come to the fore in Germany, which continued to focus on history education aimed at promoting historical awareness (*Geschichtsbewusstsein*). However, the public debate did focus to some extent on German identity as could be learned from history, although the approach to this was much more cautious than in England and the Netherlands.

Conclusions

History was introduced into teaching curricula in the nineteenth century because politicians were of the opinion that this would enhance citizenship and the formation of a national identity. The approach taken by these politicians did not follow the critically distanced standards of modern academic history, nor did it strive toward objectivity as academic historians are supposed to do.[7] "Disengaged historical interest, scientific study of the past and legitimization of the present are difficult to reconcile" (Jonker, 2001a, p. 25). According to Jonker this was a problem that had already been identified by Nietzsche in his 1874 publication *On the Use and Abuse of History for Life.* Nietzsche made distinctions within three types of history: the monumental, the antiquarian and the critical. The latter acts as a kind of counterbalance to the first two (Nietzsche, 1874/1983). However, too much of any of these three types of history is not good "for life." The dominant tendency in society and politics—as reviewed in this chapter—is to negate the critical, objective and scientific approach taken by history and to enhance the political/ideological stance in education with monumental and antiquarian history: "history taught in schools . . . responds to political imperatives" (Jordanova, 2006, p. 17). According to Jonker (2001a), "scientific work, lessons in moral, political ideas and cultural preferences are mixed up in an opaque manner" in schoolbooks (p. 25).

For a relatively short period of time—the intermezzo of the 1960s and 1970s—society and politics lost interest in traditional history education, which consequently became threatened with extinction. Paradoxically this implied new chances for developing expertise in teaching and learning history. The opportunity arose to develop ideas about the contribution that history could make to people's education, separate from societal and political pressure. Any contribution, of course, always has some political layer: this is apparent as soon as education is focused on developing "democratic citizens." However, the reformers of the 1960s and 1970s were striving toward political aims in a different way than their predecessors—aims that had not been imposed on history education through government interventions.

In Germany in this period, research was undertaken into the features of historical awareness, and people came to the conclusion that the typical historical way of orientation in time delivered an essential mode of thought for a liberal democratic society. In England thoughts inspired by taxonomies of educational objectives and the development of the new social studies from the United States were of the essence. Students' independent study was a central theme to this approach. Working with historical evidence to develop one's own image of the past was the correct approach to history teaching according to these insights. The influence of this type of educational insights in teaching and learning history has been pre-eminent in the Netherlands, particularly due to the inclusion of "structural concepts and skills" into the final examination program, which had been particularly influenced by the new social studies from the United States. In addition to the skills developed in taking a critical approach to evidence, other forms of historical reasoning were also introduced into English and Dutch history teaching. These forms of reasoning concentrated on different perspectives resulting from time differences. The introduction of the curriculum based on ten eras in the Netherlands was developed explicitly out of the intention to promote historical awareness in the way in had been advanced in German expertise in teaching and learning history.

Political and ideological tasks were once again introduced from "outside" into history teaching from the 1990s onward. Political interest for history as a school subject did not coincide with a similar focus on the development of expertise in teaching and learning history. For instance, in the Netherlands, committees such as the Committee on History Education and the Committee for the Development of the Netherlands Canon consisted primarily of members who were not experts in history education. In England, discussions about the National Curriculum were conducted by outsiders in the press. History had once again become too important to be left to teachers and educationalists alone. Perhaps this imbalance in the interests in history education and the academic expertise in teaching and learning history is not that strange, as the demands of academia and those of politics do not run parallel. The consideration that an independent, scientific approach to history with its characteristic orientation in historical time is not only the product of a liberal, democratic society, but also an essential ingredient for the development of thought in such a society, as promulgated by someone like Kocka, is not easily put across in the public debate.

The central question in this chapter was what role historical awareness has played to date in history education. The conclusion is that it has played a subordinate role. During the periods in which traditional, political and ideological aims were predominant in history education, the focus on historical awareness was limited to non-existent. The traditional, political and ideological approach to history education was, after all, not interested in emphasizing the distance between past and present, the difference in values and the contingency and capriciousness of history. Rather, it sought to emphasize shared values, a clear pattern in the development of national history destined to end up into present conditions as its natural outcome, and a shared continuity of national identity between past and present (Blaas, 1988).

This appears also to have been the case in the recent revival of interest in national history. For instance, Vos (2009) notes in his discussion of the television series *Holland in Twelve Murders* that the editors of the program wanted to kill off the myth that the Dutch had always been tolerant, level-headed and hospitable by stating that "nothing is as changeable as Dutch nature," yet they replaced one myth with another by suggesting—as is evident from these series of murders—that the Dutch had always been characterized by their inclination to reject foreign influences, by intolerance, and hysteria. The political lesson of the latter is evidently as clear as that of the former. In the television series *Holland's Past* there was a continuously recurring "us-perspective" that suggested that there had always been a Dutch nation as we now know it, which therefore existed "beyond" historical development (Van Vree, 2009).

The development of a historical awareness also continued to remain in the background in England and the Netherlands during the period in which expertise in teaching and learning history was able to work its way more or less free from government involvement. The critical study of evidence that had been influenced, in particular, by the new social studies in the United States, did not so much focus on the way evidence originated from other times and different historical contexts, but rather on social scientific questions that needed to be considered when assessing the value and reliability of information. A history curriculum based on studying evidence is unable to substantiate why the study of history should be preferred above other subjects in the social sciences. Also the present day questions and problems that were viewed in the German 1970s approach as guides to help select subject matter led to the risk of history being viewed as a "general human science," and being subsumed into social science in general.

An attempt will be made in the following chapters to approach expertise in teaching and learning history from the starting point of orientation in historical time. According to Ankersmit (1989), time is what makes history distinct from the other subjects that study social reality: "time, temporal distance is a constituent part of history, in contrast to other social sciences" (p. 12). Jordanova (2006) has noted in her general characteristics of the historical trade that history does not have its own distinct subject area, rather it is potentially concerned with all aspects of human existence; nor can history boast its own distinct research methods or theories that are able to explain change. There is only one issue that is distinctive about history: "History is the systematic study of the past, and at its heart is time" (Jordanova, 2006, p. 105). Historical thinking is primarily thinking in dimensions of historical time. If students have to learn history at school, they should also have to become proficient in developing an awareness of historical time.

2

Historical Time

What is historical time? What are the characteristics of a historical consciousness of time, and what focus points do these features provide for the development of expertise in teaching and learning history? In order to answer these questions, this chapter provides a summary of the main points in the philosophy and theory of history about time and historical time. Historical time will then be elaborated in terms of central concepts that could be of interest to the expert on teaching and learning history. Potential objectives for history teaching follow from this, alongside the associated research questions for academic expertise in teaching and learning history.

The description in this chapter is aimed at finding a workable definition of historical time that can be used in teaching and learning history. To be able to arrive at such a definition, it is unavoidable to pay some attention to the most important philosophical notions about time in general. The specific features of a *historical* consciousness of time can only be outlined against this background. Therefore a number of key documents are used in this chapter to firstly discuss the primary problem areas around the concept of time in general, with the aim of subsequently developing a working definition of the concept of "historical time consciousness."

Images of Time, pages 33–93

Time is an intangible and elusive concept. On the one hand time seems to be given as a self-evident natural phenomenon, something that as such does not require much attention. On the other hand the subjective experience of time, the way in which we perceive time and the passage of time, is essential for our thoughts about time. Time has "objective" and "subjective" dimensions. It is necessary to focus on both of these aspects in order to get a good understanding of the issues involved in studying time. Historical time includes both the objective and the subjective approach of time, as will be demonstrated below. This chapter has therefore been structured as follows: firstly there is a discussion of aspects of objective and subjective time (Sections 1 and 2), in order to subsequently be able to discuss the specifics of historical time consciousness (Section 3). This specific character is then accentuated by drawing a comparison between the historical and other possible temporal experiences (Section 4). In the penultimate section of this chapter (Section 5), the findings about historical time consciousness are applied to expertise in teaching and learning history. A final overview is provided in the last section (Section 6).

Objective Time

Objective time is understood to mean time as it is given as a natural phenomenon. Because the passage of this time is clearly embodied by the movement of heavenly bodies, it is also referred to as cosmological time. In the work of Ricoeur this type of time is referred to as "objective time," "time of the world," "ordinary time," "star time," or "universal time" (Ricoeur, 1988, pp. 104, 107). These terms seem to refer to the same thing; at least, there is no explicit distinction among these terms provided in Ricoeur's work. There are two types of temporal experience out of eight distinguished by the anthropologist Hall that fall into the "objective" category: physical time and biological time, which he summarizes with the terms "explicit, technical, existential" and "low context" (Hall, 1983, pp. 16–19, 20–22).[1] The latter is intended to refer to the types of time that have little association with the cultural contexts in which people use them.

The movement of heavenly bodies, as stated, gives shape to this time. The revolution of the earth around the sun determines a "year"; the revolution of the earth around its own axis determines a "day"; the movement of the moon around the earth determines a "month" in principle and causes the ebb and flow of the seas.[2] People, animals and plants display a biological life rhythm through their existence in this time, which, as the global travelers among us will know, can be disrupted thoroughly by a rapid displacement from one time zone to another. The manifestations of natural time are thus enumerated exhaustively. All other divisions of time have been added by humans: the distinction of hours, minutes and seconds; the arrangement of days in weeks; the distinction among four seasons; as well as the start and the end of the year. All of these apparently objective instances are products of human culture, which could have been devised differently. For instance, the Mayans and Aztecs based their calendars on a twenty day period in stead of a seven day week; this was probably chosen due to the number of fingers and toes on a human body (Aveni, 2002; Boot, 1989). The Balinese use "weekly cycles" varying from two to ten days that overlap with each other, in a way similar to how weekly and monthly cycles overlap in western culture (Howe, 1981). The ancient Egyptians divided their year into three, rather than four seasons (Parker, 1950), while the Saulteaux Indians along the Berens River in North Manitoba (Canada) distinguish six seasons (Hallowell, 1937). However culturally determined—and in certain respects awkward[3]—our normal time division may be, it is the only measurement system that has not become streamlined by decimalization, although attempts were made to do this during the French Revolution (Bianchi, 1982; Zerubavel, 1977). Against the background of the success of other types of decimalization, the failure to decimalize time perhaps demonstrates how much we are fused in with our experience of time: tampering with the cultural tradition in this respect affects the core of our existence.

When including further divisions of time beyond year, day and month, instead of just natural time we have to speak of calendar time and clock time, which strictly speaking are not given nor objective (in Hall's categorization they form "profane time"). Nevertheless, they will be discussed in this section on objective time, as the passage of this type of time occurs to us as a natural law. To put it more boldly: many natural laws can only be determined on the basis of clock time. The relationship between objective time and important natural laws, such as Newton's first and second laws, is a fundamental basis of thought in natural science (Le Poidevin, 2003).[4] Ankersmit (1989) refers to clock time as being "entirely unproblematical from a philosophical point of view" (p. 16). This does not detract from the fact that many philosophers have struggled to comprehend objective time. Three problems associated with this kind of time will be discussed in sequence in the next subsections: the existence or non-existence of time, the relationship between objective time and the subjective observer, and the relationship between objective time and movement or change.

The Existence of Time

Time was part of Aristotle's discussion in his *Physics*, which makes clear that he believed time to be part of the natural world (Aristotle, 2001). The concept of time is a prerequisite for the physicist's work, as is the concept of space, which is also dealt with in Book IV of the *Physics*; everything takes place within space and time. The first problem facing Aristotle was whether time actually exists. After all, some part of time has passed and is not part of the present, while another part lies in the future and has not yet occurred. Yet both the past and the present belong to time, as it is difficult to maintain that time only exists in the "now." This gives rise to the question of whether something that consists of the non-present can be said to belong to the category of things that exist. Viewing time as a series of "nows" is not a solution, as this gives rise to the question of what the borders of that kind of "now" are, while "now" is also a border between the past and present. It is impossible for a "now" to be bordered by two other "nows," which would all represent the present. Thus, the existence or non-existence of time remains an intractable problem (Aristotle, 2001).

This matter continued to intrigue philosophers for centuries. For instance Newton also wrestled with the concept of time, which played an essential role in his natural laws. He had to take refuge in metaphysics in order to be able to postulate an "absolute" time running with arithmetic regularity. According to Newton time is neither a substance nor an event, but it has its own manner of existence "emanating from God" (Turetzky, 1998, p. 72).[5] The existence of God implies the existence of time, as according to Newton anything that exists has to exist at a given time. Therefore time, as God, must be defined as unending and not part of the Creation. There is an absolute time or "duration" that is independent of any other phenomenon: "Absolute, true, mathematical time, of itself and by its own nature, streams forward constantly without any relationship to anything beyond itself, and is also called duration" (Newton, 1687, p. 11).[6] Newton was unable to prove the existence of time. He took it as an axiom and then built his system of natural laws based on that point of departure.

The British philosopher McTaggart proposed the opposite to Newton, namely that time does not exist (Le Poidevin, 2003; McTaggart, 1908; Turetzky, 1998). According to McTaggart a position in time is determined by the fact that it is earlier or later than another position, as well as by the fact that it belongs to the past, present or future. One of the definitions of position changes constantly as events belonging initially to the future subsequently become the present and then the past. The other does not change: something is earlier or later than something else, and that will always be the case. McTaggart referred to both series in time as the "A-series" and the "B-series," respectively the changeable one of past, present and future and the immutable one of earlier and later. The A-series is, according to McTaggart, more fundamental for time than the B-series, as the idea of the passage of time is expressed in the A-series in terms of changeability. Besides the B-series does not allow us to ascertain when "now" occurs; it only determines relative positions in time. The existence of the A-series is therefore a prerequisite for the existence of time. According to the A-series all phenomena in time have the properties "past," "present," and "future." This leads to an absurdity unless we say that they do not have these properties concurrently, but rather successively. Yet in order to be able to state this, we have to use the concept of time, which is defined in turn by the A-series. This would lead to a circular argument. The A-series can therefore not exist, and it follows that time is something unreal. McTaggart ascribes the fact that we nevertheless reason in terms of past, present and future to the human mind, which is set up in such a way that observations in the present are experienced as "now," memories as "past," and expectations as "future." However, this kind of conceptualization of time rests on a misunderstanding of what is, in fact, a timeless reality.

The discussion of McTaggart's paradox led to the creation of "A-theories" and "B-theories" in the twentieth century. The first are based on the assumption that only the A-series represent real time, but attempt to solve the problems envisaged by McTaggart within this. The second are based on the assumption that only relationships in the B-series are true temporal relationships and that there is therefore no paradox (Turetzky, 1998). An example of an A-theory was provided by Broad, who believed that a linguistic issue was at stake. By saying that a temporal phenomenon is always past, as well as present and future, one is defining temporal phenomena by linking them via a non-temporal association ("is") to an adjective that suggests that a temporal phenomenon has certain characteristics that conflict with its other features. However, perhaps we should not be thinking at all in terms of characteristics and therefore avoid making use of adjectives with these non-temporal links. Perhaps we should only be speaking in terms of "was," "is," and "will be" (Broad, 1983; Turetzky, 1998).

The entry in the *Stanford Encyclopedia of Philosophy* for "time" proves that McTaggart uncovered a real philosophical problem about which we have not yet heard the final word. The discussion of McTaggart's paradox plays a prominent role in this entry (SEP, 2002; Carter, 2003). It is clear that the existence of objective time is by itself not an obvious statement in the philosophy of science. Newton's claim of an evenly elapsing time was generally accepted by science until Einstein proved that time is relative. Nevertheless, in our daily existence we assume that there exists an evenly elapsing Newtonian time. The non-existence of time is not something that we can easily accept, since all our experiences suggest the opposite. If time is the determining factor in history, and time does not exist, this would mean that history is based on something that does not exist, which is not easy to accept either. However, up to now the existence of objective time is philosophically an *aporia*, an insoluble problem.

The Relationship Between Objective Time and the Observer

McTaggart's theory pointed in the direction of a subjective interpretation of the concept of time: time is perhaps not an objective phenomenon, but a construction of the human mind. Kant had previously indicated the importance of the observing subject in relation to the concept of time. Kant's philosophy concentrated on the question of how knowledge is possible through the association between empiricism and categories that are *a priori* present in the human mind. Time, for Kant, is an *a priori* concept that lies at the basis of and enables our observations. Time is not an empirical concept, as without time no observations could take place (Kant, 1787). We would not be able to determine whether things exist concurrently or successively. However, if one bases the argument on objects "by themselves" and moves away entirely from human observations, then there is no time (Kant, 1787). Time is therefore bound to the observer: "time is solely a subjective condition of our (human) observation . . . and by itself, outside of the observing subject, nothing" (Kant, 1787, par. 6: c, my transl). However, time is also absolute as a prerequisite and an *a priori* condition to the possibility of knowledge. The necessity of time for observations translates into a number of axioms: time has only one dimension and different times do not exist concurrently, rather they exist consecutively (yet different spaces do exist concurrently). Therefore time should not be envisaged as a line (parts of which exist concurrently), but rather as an arithmetic series. These are the rules that make our observations possible (Kant, 1787).

Aristotle had already suggested a relationship between time and human observations. He cites a legend of men who had slept alongside heroes in Sardinia for a hundred years without their bodies undergoing any change during that time. When they awoke, they continued their lives as if it was the next day and not a hundred years later (Aristotle, 2001). The fact that they had not *noticed* that time had passed, made them assume that time *had not* passed. Aristotle concludes:

> One might ask whether time would be or not be if there were no soul. After all, if it is impossible for someone to count, then it is impossible for something to be counted, and it stands to reason that in that case there is also no number. (2001, Book IV, p. 14)

Time therefore requires an observer in order to exist. Strictly speaking this type of reasoning is not entirely correct, as an epistemological argument is being used for an ontological conclusion: something unknowable cannot exist, and something exists because it is being observed. Aristotle would normally reject this type of reasoning; however he seems to believe that the argument is strong enough in the case of time (Turetzky, 1998).

Nevertheless there seems to be something within us that tells us that time would continue to advance even if there were no people around to observe it. The passage of geological time and time during the evolution of life before humans were around as observers does indeed suggest that this is the case. This is the second insoluble problem in the philosophy regarding objective time, which must also have frustrated Aristotle and Kant. After all, they were attempting to describe how "objective" knowledge of the natural reality was possible. Their statement that time could not exist without an observer does not mean that they were attempting to arrive at a subjectivating interpretation of the concept of time. They were concerned with knowledge of nature, not human experience.

Time and Movement or Change

Aristotle establishes a relationship with movement when talking about the nature of time. He notes that others had already established this before him. One of those could have been Plato, who describes how the world and therefore time originated in his *Timaeus* (n.d.): as a moving image of eternity (of eternal forms or ideas), with a fixed, numerical movement, divided into months, days and years, and also into past, present and future representing types of time. Time was created at the same moment as heaven (and heavenly bodies) and would disappear when these to came to an end, as the movement of the heavenly bodies was created in order to be able to observe the number of time (Plato, n.d., Book I). However, Aristotle found that equating time with movement was not a satisfactory argument. After all, movements may be faster or slower, but time cannot change speed, and without time, a change of speed could not be noticed. Time is therefore "something of the movement" (*ti tès kinèseos*) (as cited in Ricoeur, 1988, p. 13). Time, like movement, has something to do with "earlier" and "later," but earlier and later are not in the same way features of time as of movement. Time, after all, is the measure of movement, which maps out and sets the limits for movement. What is being limited cannot be the same as what is setting the limits, and a number cannot be the same as that what it is counting. "It is clear, then, that time is the number of movement in respect of the before and after, and is continuous since it is an attribute of what is continuous" (Aristotle, 2001, Book IV, p. 11).

Kant established the relationship between time and movement or change in a different way from Aristotle. According to Kant (1787), time is not an aspect of movement or change, but precedes it in the sense that perceptions of movement or change are only possible because of time. It is only due to time that something is present as well as not present at a certain location; that is, time makes these two conflicting statements about a certain observation compatible. Therefore, Hegel put an emphasis on "becoming" in his observations about time: everything in the world comes into being and disappears. The past is the actual existence of the world that has ceased to exist and the future is the non-existence of the world that will once be. The present unites the past and future in a negative sense, as the present comes into being because the past ceases to exist, while the present exists now as its non-existence lies in the future. Time is the contradiction in and continuous movement of finite beings (Turetzky, 1998).

The relationship between time and movement is made clear by imagining a "temporal vacuum": a time period in which nothing moves or changes in the universe, including thoughts and human states of consciousness (Le Poidevin, 2003). This thought experiment leads to logical contradictions in a number of ways. For instance, the length of the temporal vacuum could not be established in any way, while every time period has a certain duration.[7] It could not be excluded that a temporal vacuum of say a hundred years occurs between the moment at which the reader reads this sentence, and the moment at which he reads the next in which nothing moves or changes. Although this is imaginable, it is, of course, also an absurdity to imagine this. This leads to the conclusion that the passage of time is inextricably linked to change and movement, and yet time is not the same as movement or change. This is a third insoluble problem in the philosophy regarding objective time.

Conclusion

Time is perhaps not as "objective," "given," and unproblematic as one usually thinks it is. The existence of time constantly seems to have to be linked necessarily in one way or another with human beings as observers. Time is not the same as change or movement, yet without change or movement it is virtually impossible to imagine the passage of time. A problem with the metaphor of space—which is so readily used in relation to time, for instance, through the use of timelines or talking about "great distances in time"—is that things may exist concurrently in space, but not in time. Space has three dimensions, yet time only has one. Thinking about time leads to complications that at first glance appear to be trivial and irrelevant puzzles, yet lead after further consideration to serious philosophical conundrums. Perhaps these are partly cause of the fact that it is difficult to learn to think in terms of passage of time.

Subjective Time

Subjective time is understood to mean time as experienced by human beings. Temporal experiences may occur in various ways to people. The anthropologist Hall distinguishes at least five categories of time that may be defined as subjective (Hall, 1983). It became clear in the above discussion that human observation is an essential aspect of time, including "objective" time. So, unsurprisingly, there is a school of philosophical thought about time that takes human experience of time as its point of departure. The recognition that time is inseparably linked to human experience led these thinkers to assume that a proper description of time could only be derived from human experience. The way towards this school of thought was opened up by the church father St. Augustine.

St. Augustine on Time

Skeptical arguments giving grounds for doubt about the existence of time were not acceptable to St. Augustine. Time, after all, belonged to God's creation. The eleventh book of his *Confessions* is intended as an explanation of the creation narrative from Genesis 1. In this initial passage of the Bible, time is contrasted with eternity, which led St. Augustine to the question "What then is time?" He argues against skepticism by departing from human experience in which time is a familiar, daily phenomenon: "Yet is it not true that in conversation we refer to nothing more familiarly or knowingly than time?" (Augustine, 1992, Book XI, 14:17).[8] St Augustine makes use repeatedly of the stylistic feature "and yet..." (et tamen) when creating a counterargument against reasoning out of which it would follow that time is something unknowable and immeasurable: and yet we refer to a long or short period of time, and yet we observe time intervals (Augustine, 1992, Book XI, 15:18, Book XI, 16:21, Book XI, 27:34).[9] The fact that we do this demonstrates that people are able to observe and measure time, but how? The past and future must somehow be there in the present, because if the past would truly be past, then it would no longer be in existence, and if the future would truly be the future, then it would not yet be in existence (Book XI, 18:23). This brings him to the conclusion that three times are "present": a present of the past, a present of the present, and a present of the future (Book XI, 20:26).[10] Where are these three present? St. Augustine states that they are present in the soul—for "I do not see them anywhere else"—namely the past as memory (*memoria*), the present as observation (*contuitus*), and the future as expectation (*expectatio*) (Book XI, 20:26).[11] As has been explained in the previous section, McTaggart (1908) came to a similar insight at the start of the twentieth century, although the way in which he arrived at this was completely different from that of St. Augustine.

St. Augustine develops his findings in more detail by focusing on speaking and listening, in which the stream of words and letters forms a single entity in our experience, yet still belongs partly to the past, partly to the present and partly to the future. Something similar also happens when reciting a poem that we observe as a single entity in the process, although there is a part that has passed, a part that we are presently reciting, and a part that is still to come (Book XI, 26:33–27:34). The conclusion of this line of reasoning is that time is nothing other than a distention, but of what? "It would surprise me if it were not a (*distentio animi*) distention of the soul itself" (Book XI, 26:33).[12] This distention of the soul is not just that of the observer, of a passive receiver of impressions about the past, present and future. A singer singing a long note knows in advance how long the note will be and is aware when singing the note for how long it has already been sung and how long he has still to maintain it. So the note is completed while the present intention transfers the future into the past (Book XI, 27:36).[13] Temporal experience therefore concerns *intentio* as well as *distentio animi*.

The intentional actions are being defined here next to and within the "distention of the soul" as constitutive elements of temporal experience, through which past, present and future are no longer three separate worlds, but rather a single coherent entity defining this experience. Thus, St. Augustine's argument opened up the way to a phenomenological conceptualization of the phenomenon of time.[14] Attempts to conceptualize time as an objective natural phenomenon constantly lead to doubts about the existence of time or to logical inconsistencies. Time is being lived through and experienced by human beings; time is an essential component of human existence in which memories and expectations determine actions continuously. Time is difficult to define outside of human existence and human experience.

Phenomenological Conceptualizations of Time

It took until around 1900 for these approaches to time to be further developed in philosophy. In the intervening centuries a sharp distinction had been made in western philosophy between object and subject (Turetzky, 1998), which meant that subjectivating approaches to the phenomenon of time became far less probable. The Kantian turn signaled an end to this, although Kant's statement that time does not exist outside of the observing subject did not lead him to make temporal experience by the subject the focus of his work. He merely determined which *a priori* categories played a role in a knowing subject when observing the natural world around him as clearly as possible.

Bergson (1888) developed a theory about time that was close to St. Augustine's train of thought. Like St. Augustine, he explained temporal experiences using the way in which we perceive tones and music. Just like St. Augustine warned against equating time with movement, Bergson (1888) warned against confusing space and time, although both are an "empty homogeneous medium" (p. 47), which at first glance would lead one to believe that only one of these could exist. This, according to Bergson, is the reason why we confuse time with space as soon as we consciously attempt to create a representation of it: we then see time in its extensiveness as if past, present and future were all aspects of the present, and we imagine time as a line. The drawback of this kind of conceptualization had already been brought to the fore by Kant (1787). After all, past, present and future do not exist at the same time. This simple observation contains a warning against the all too easy use of spatial metaphors for time, according to Bergson (1888): "this simple reflection should warn us that unconsciously we thus revert to space" (p. 47, my transl.). By thinking in terms of a time line, we return to spatial categories without realizing it.

An experience of "pure duration" provides a better idea of what time is, according to Bergson. This type of experience occurs in its purest form when people do not consciously imagine a passage of time, but rather get totally absorbed in time:

> The completely pure duration is the form which is taken by the succession of our states of mind when our 'me' lets itself live, when it renounces attempts to establish a separation between the present state of affairs and previous states of affairs. (Bergson, 1888, p. 48, my transl.)

Something like this occurs, for instance, when one listens to music: dwelling upon each note separately, or observing a note in a particular way would disrupt the appreciation of the musical composition as a whole. One needs to let the whole wash over oneself in its stream from the past via the present to the future and experience it as a single entity. This reasoning has close resonance with St. Augustine's example of reciting a poem, although

Bergson places more emphasis on experiencing the pure duration and less emphasis on the "intention." He seems to believes that pure duration can also be abstracted from our experience of constantly changing matters, just like a completely empty space is an imaginable category. That would imply that pure duration is not a subjective category.

The founder of phenomenology, Husserl, viewed time as the "basic form of all experience" (Dostal, 1993, p. 145). He attempted to fathom the nature of time by "phenomenological reduction," that is—true to the principles of phenomenology—by "bracketing" (*Einklammerung*) objective time that obeys natural laws, and starting exclusively from the way time occurs to human beings. In his *Vorlesungen zur Phänomenologie des inneren Zeitbewusstseins* (Lectures on the Phenomenology of Inner Time Consciousness), Husserl states that time manifests itself in the first instance in the "now," which needs to be interpreted as a "thickened" now that carries with it "retentive" and "protentive" aspects (Husserl, 1928, pp. 30–32). Each "now" is determined in a person's experience by what has been retained from the past and by what is expected for the future: the retentions are bound to each now-moment like a "comet's trail" (Husserl, 1928, p. 25). The relationship with St. Augustine's threefold present is clear, but Husserl adds something important to this.[15] According to Husserl, while time continues on, each moment is retained in the following moment; retention and protention unconsciously determine our daily experience, which occurs as a "stream of experiences" in a single irreversible direction. Retention and protention are therefore not the same as "memory" and "expectation." The issue at stake is the self-evident temporal experience that is always present in everything that occurs to us. This experience may be compared to the background and the horizon against which we place observations of every spatial object. Although we do not concentrate on them, we do notice the background and the horizon (Husserl, 1928; Carr, 1986).

Just as we can consciously imagine spaces outside of our observable horizon, we can also consciously form images of the past and future not belonging to the "thickened" now. Such a conscious recall of a past that does not belong to the "now-experience" is a conscious "reproduction" of a now-moment from the past, including the retentions and protentions that were then associated with that moment; Husserl (1928) refers to this as "secondary recall," in contrast to primary retentive recall that is included with each now-moment. Consciously developed images of the future are not protentions either, but secondary future expectations. Secondary recall and secondary future expectations come and go; sometimes they are emphatically present and sometimes not, yet retention and protention are always there, and belong to each "now." That is not to say that retention and protention only relate to the moment just past or the immediately expected future. Decisive is not proximity or distance in time, but the implicit way in which the past and future are enclosed in the present.

Husserl's student Heidegger (1927/1993) explored these ideas further. The aim of his principal work, *Sein und Zeit* (Being and Time), published in 1927, was not to create a phenomenology, but rather an ontology of time. He nevertheless applied a phenomenological method to arrive at this. In order to grasp the essence of "being," Heidegger departed from "being there" (*Dasein*), namely the being there (existence) of human consciousness, the only entity that is at all concerned with "being." In fact, Heidegger meant *Dasein* to refer to "man" or "a human." In the following section, for the sake of clarity, the wording "human being" is used where in the original text we find "being there." According to Heidegger a human being is temporal through and through; human beings exist in time and are even a *sein-zum-Tode* (being towards death), as death is the certain perspective of every human being (Heidegger, 1927/1993, p. 234). This means that all human perceptions and experiences are bound to time. Human beings are *innerzeitig* (within time) and *geschichtlich* (historical): man exists in time and in the midst of circumstances created by a historical development, which have not been chosen or shaped by him—they are simply there. The existence of a human being at a given moment and under certain circumstances does not have any *a priori* sense or meaning—that is to say, no sense or meaning which are known to that human being. Human beings just happen to have arrived somewhere in time in the midst of certain circumstances, a condition which Heidegger (1927/1993) refers to as *Geworfenheit* (having been thrown) (p. 136). Humans have to determine their attitudes towards the world and history. This may occur in a "proper" or "improper" way. The latter occurs when humans allow time and circumstances more or less to determine their lives without deliberately choosing a particular position vis-à-vis their circumstances, thus lapsing into "the quotidian." However, the "proper" human being is characterized by a deliberate approach to the circumstances created by tradition and history. According to Carr this is an important distinction: "the difference is between the blind and unreflecting follower and the proud and conscious bearer of a tradition" (Carr, 1986, p. 94).

Departing from "being there" Heidegger hoped to be able to ultimately discover the relationship between being and time; however this work remained unfinished. What we are left with is the relationship between "being there" and time as described in the completed sections of *Sein und Zeit*. This relationship is characterized among others by *Sorge* (care), that is the way a human being relates to the world and deals with it. This is associated with time in three ways: by “being ahead of oneself” (*sich vorweg sein*); by “already being in,” which tallies with *Geworfenheit*; and by the everyday “being with” the world and the people who happen to be around us (Heidegger, 1927/1993, par. 50, pp. 249–250). These three are related to a forward look towards the future, to finding oneself in the present as it happens to have arisen without one’s own choices, and to the daily existence in the present. In this way Heidegger also arrived at something like the “thickened now” of his teacher Husserl, or something that seemed to be similar to the threefold present of St. Augustine.

A similar kind of threefold present is found in Oakeshott’s work. For instance, in his *On History* (1983) he describes an example of a pedestrian on the edge of the sidewalk at the point of crossing over and another who has a wooden leg. In the first case a future event can be drawn out from the observation; the second instance refers to something that has occurred in the past. The future is here an understanding of the present in terms of a change it appears to be suggesting, and the past is an understanding of the present in terms of a change that it appears to have retained or maintained. This brings to mind Husserl’s retention and protention, although here it concerns the observation of the attitude or status of *someone else* in which past or future have been included, and not the personal experience of the passage of time as takes place when listening to music or reciting a poem. Oakeshott (1983) refers to this as a “practical present-past” and a “present-future of practical engagement” that needs to be distinguished from a historical past (p. 19). What he sees as a historical past is discussed in the next section on historical time.

Conclusion

Phenomenological thinkers departed from man’s temporal awareness and deliberately rejected attempts to define time as an objective natural phenomenon outside of human experience. Yet Bergson refers to “pure duration” as something that stands by itself. Ricoeur (1988) points out that the phenomenologist is forced to admit that there is a relationship between the “stream of consciousness” he is studying and the “objective passage of time,” just as there is a relation between “the one following the other” in immanent time and the succession of events in objective time (p. 24). Phenomenological time can only be articulated by utilizing units derived from objective time. On the other hand Newtonian “objective” time can only be experienced via a subject’s activity. The principles of phenomenology and critical (Kantian) philosophy are mutually exclusive, yet they do borrow ideas from each other (Ricoeur, 1988). At the end of the previous section of this chapter about objective time, it had to be concluded that such a time perhaps does not exist by itself. In a similar way it has to be stated here that a comprehensively "reduced" phenomenological perception of subjective time does probably not belong to the realm of possibilities either. It seems that objectivating and subjectivating perspectives are condemned to each other in thinking about time. This is expressed, for instance, in the observation of people growing older that time seems to run faster as their age is advancing: the speed at which they experience new impressions and changes in their lives decreases, while subjective temporal experience is "accustomed" to measuring time by the tempo of events sequencing each other; consequently objective time seems to be moving faster (Roeckelein, 2008; Draaisma, 2006).

A distinction that is relevant for the practice of history (and history teaching) is the difference highlighted by Husserl between retention and secondary recall. This indicates an important difference in temporal consciousness as such and historical time consciousness, which—in Husserl's terms—is always based on secondary recall. The distinction drawn by Oakeshott (1983) between the "historical past" and the "practical present-past" suggests something similar. Apparently the consciousness of a historical time has a separate place among the ways in which people experience time.[16] Historical time consciousness is not automatically included in every "now-moment," as does appear to be the case for temporal experience in general. Historical time consciousness involves a deliberate choice and is therefore also one that can be ignored. Expressed in educational terms this means that historical time consciousness needs to be learned. The specific character of historical time consciousness is discussed in more detail in the following section.

Historical Time

Historical time is the experience of time such as it comes to the fore in the historian's work. The dimension of time is essential for history, but that does not mean that history alone is concerned with time. Other subjects also deal with it. So it has to be the way in which time is considered that distinguishes between historical time and other temporal experiences. Sciences such as geology, astronomy, and (evolutionary) biology study developments that have taken place in nature over very long periods of time. However, history is concerned with human culture and therefore the intentional shaping of both the developments and events described as well as of those descriptions themselves, and it is therefore also concerned with the intentional conceptualization of temporal experience. History concerns the passage of time on a human scale, not on a scale of the universe, the development of the earth's crust or the development of life on earth. Both the subject and the object of historical time are to be found within a cultural environment.

The latter also applies to social sciences such as sociology, anthropology and political science. History is traditionally seen as being distinct from these because it is an "idiographic" science focused on the particulars of each case, while social sciences are focusing on general laws and as such are to be considered "nomothetic," although it is difficult to draw clear boundaries between the two (Ankersmit, 1984; Lorenz, 1987; Van der Dussen, 1988). General laws about human societies that transcend the particulars of each case and each time appear particularly difficult to find, while the focus on the particulars of each individual case or each different time period could never exist without general categories playing a role. Nevertheless the interest of the historian in the unique is undeniable, and the absence of a tendency to highlight general laws is generally considered to be characteristic of history (Evans, 1997). This has its consequences for the way in which historians deal with time: it is not the logical and continuous line of developments that draws their attention, but the capricious, the constantly changing and unpredictable. Insofar as change and movement are characteristic of temporal experiences, one could therefore say that the historian places more emphasis on temporality than the social scientist does. Jonker (2001a) refers to the "tenets of temporality and contingency" (p. 139) through which the historian has to limit himself by necessity to retrospective knowledge, and cannot be tempted to provide recipes for (future) solutions of social evils, as is the social scientist's lot. Historical insights, on the contrary, reveal the relative value of these recipes.

There is some logic in the fact that historical time has characteristics that combine both types of temporal experiences described in the first two sections of this chapter. The objectivating temporal experience tends towards measurement, and the subjectivating towards experience; both play a role in history. On the one hand history involves images of the passage of time as sketched out by humans, for instance by distinguishing between "periods" with their own characteristics. On the other hand historical time is bound to more or less objective, chronological laws. This can be explained in the simplest terms as follows. According to the historical periodization approach to the past, there has been a period called Middle Ages which ended at a certain moment in time to turn into the Early Modern Period. However, objective chronology suggests that such a distinction is artificial: there was merely a succession of years, as there had always been. These periods did not exist for the people who experienced them; this periodization has only come about through a retrospective look at images of the past.

A historian *has* to develop images of time, including characterizations of eras, such as "Middle Ages," because merely enumerating events within a chronicle or list of years does not create history. However, the succession of events and phenomena in line with objective chronology is an external measure the historian cannot ignore. The objective and subjective are therefore emphatically linked to each other in historical time. The following sections describe how this translates into a number of characteristics of historical time.

The Development of Western Historical Consciousness

The manner in which we perceive historical time is connected with developments in Western culture over the last three centuries. The British historian John Pocock set out the conditions that can give rise to historical consciousness in an article published in 1962, which has afterwards been described by Blaas (1988) as "fundamental" (p. 4). In addition to the availability of historical evidence and concepts in which the past can be described as past, according to Pocock (1962) an important pre-condition for the emergence of historical consciousness is a situation in which the relationship between the past and present is viewed as being problematical and in which there is a sense of discontinuity: "the mere affirmation of continuity can produce only traditionalism: historical explanation can arise only where there is some awareness of discontinuity" (p. 237). Where the past and present are viewed as a continuous entity, the study of the past is felt as superfluous, as only "more of the same" could be encountered in the past. The same traditions are observed in the past as are still in existence in the present. Only a sense of a breach will give rise to the thought that the past was quite different from the present. This awareness developed in Western culture in association with the scientific, technical, industrial and political developments from the second half of the eighteenth century onwards (Carr, 1986). The industrial and democratic revolutions changed the outlook of the world so significantly and at such a rapid pace that the ties between the past and present were disrupted.

Up to that point the past could be considered as not being essentially different from the present, thus creating a reservoir of images that could be used as examples for mankind in the present: *historia magistra vitae* (Koselleck, 2004). So much continuity was experienced between the past and the present that the past could be easily used as a model for the correct forms of behavior in the present. This feeling came to an end through the turbulent developments of the scientific revolution, the democratic revolution, and the industrial revolution. The *Querelle des Anciens et des*

Modernes—a polemic conducted at the *Académie Française* at the end of the seventeenth century about the question of whether the Ancient World had been the perfect model for all time or whether the Ancients could be surpassed by the Moderns—marked the prelude to a new temporal consciousness (Blaas, 1988; Lowenthal, 1985). The observation that the Modern Times could do better than the Ancient led to the latter becoming no more than a historical period from the past. Thenceforth the idea of development and progress could take root, and therefore of "other times" that were so essentially different from the present that a comparison was no longer sensible or instructive to daily practices in the present. For this reason, the rise of modern historical consciousness is sometimes located in the sixteenth century by some, as the Renaissance and the Reformation also created a significant sense of a breach with the past. However, locating this breach in the second half of the eighteenth century appears more plausible, if only because a modern historical science came about partly as a result of this breach during the nineteenth century, which had not been the case earlier (Lowenthal, 1985).

The enormous "acceleration" of historical developments through the democratic and industrial revolutions meant that the world of the past was put at a distance (Koselleck, 2004). "The past evidently was a foreign country. Its contrasts might amuse, shock, or even instruct the present but could no longer sustain faith in its exemplary role" (Lowenthal, 1985, p. 102). Gadamer (1987) refers to this revolution as "very likely the most important revolution among those we have undergone since the beginning of the modern epoch" (p. 89). De Certeau (1988) observes that modern history was made possible through the distinction between the past and present, by which the past was positioned vis-à-vis the researcher as an object of study in a similar way that knowledge about the human body only became possible when the body became viewed as no more than a dead corpse to be dissected in order to increase our understanding of it. The growing tension between what Koselleck (2004) calls the "space of experience" and the "horizon of expectation" determined modern historical consciousness: in the traditional world of farmers and tradesmen, experience taught that little or nothing would change, but the modern world revealed that expectation patterns needed to be radically adjusted. The past and present had grown decisively apart: other times would be essentially different thenceforth.

Ankersmit (2007) characterizes the breach with the past as occurred in the Western world as a "traumatic experience" (pp. 357, 387). Western man was forced to enter a new world, yet was at the same time aware of a previous world that was no longer his own. This awareness became part of his new identity: he knew he had to become what he was no longer. A return to the past was no longer possible, yet the yearning for the old world translated into a craving for knowledge of it: *"Was einst Jubel und Jammer war, muss nun Erkenntnis werden"* (What used to be triumph and sorrow now has to turn into insight) (Burckhardt, as cited in Ankersmit, 2007, p. 361). The insight into other times that followed from the objective study of the past led to each time period being considered by itself, and not just as an antecedent to successive periods or the present. "Every era exists in a direct relation to God, and its value rests not at all on developments ensuing from it, but on its own existence, on its own self" (Ranke, as cited in Picard, 1927, p. 237).[17] This became the motto of nineteenth century historicism. Each period of time had to be studied, understood and evaluated by itself. In this way history rather increased the sense of distance to the past than reduced it.

The demand for each period to be understood by itself also created a paradox. How should a historian properly understand the thoughts and observations from a different time, if these were really essentially different to the thoughts and observations from the present? German "hermeneutic scholars" such as Droysen and Dilthey believed that this was possible based on the capability of human beings to empathize and sympathize (*verstehen*) with other human beings (Ankersmit, 1984).[18] But if this was really the basis of historical knowledge, that would mean that there existed something generally human transcending different times—something which was denied by historicists. That is, their attention was drawn to the difference between past and present, rather than the communal (Ankersmit, 1984). How could essentially different aspects of earlier times be acknowledged and apprehended by people from later times, without the personality of the "translator" playing a role? It is clear that the historian who is the creator of images of the past would have to be related to the theory of historiography sooner or later.

The rise of historical thought meant that Western man thenceforth would need to distinguish a series of "times," each of them essentially different from the others and with its own characteristics: *periodization.* Something belonging to a particular time was not to be confused with another time. A sense of *anachronism* is therefore also a feature of modern historical consciousness (Blaas, 1988). This consciousness is associated with the unpredictability and capriciousness of historical developments: after all, each time stands alone and is not to be regarded simply as the antecedent to subsequent times. Each period of time carries along eventualities that later on appear to be realized, as well as eventualities that later on appear not to be realized. Therefore a direct line cannot not be drawn from the past to the present, as this would lead to a lack of interest in eventualities that were not to be realized later on, as well as an over-appreciation of those that were to be realized, which would do no justice to the time period in itself. Thus, a focus developed on the unpredictability and coincidence (*contingency*) of historical developments.

Three key concepts of historical time consciousness can be drawn from the modern historical consciousness as has developed over the last three centuries:

- periodization,
- anachronism, and
- contingency.

These three concepts will be further developed and explained later in this chapter.

Emplotments of Time

The "paradox of hermeneutics" has already clarified that the historian and the way in which he shapes images of the past cannot be left out of consideration, even though historiography has an authentic interest in "doing justice" to the past as it "really was." Since the appearance of Hayden White's *Metahistory* (1973), the mold in which the historian pours his image of the past has received a lot of attention in the philosophy of history (Ankersmit, 2007), which meant taking a radical step away from historicism. White (1973) renounced the idea that the focus should be on the events themselves and took the position that history only existed as a text, as a "verbal structure in the form of narrative prose" (p. 2), intended as a model or icon representing structures and processes from the past and thus explaining them. The historian firstly constructs stories from the "unprocessed historical record" by viewing events as the beginning, middle or end of a certain development. The historian then "emplots" these stories in such a way that they acquire a certain meaning. The historian therefore does not "discover" or "expose" stories from the past: he *creates* the stories and emplotments via the literary form he chooses for his historical narrative (White, 1973). The relationship between the objective passage of time—as reflected in a chronicle—and the subjective form in which the historian creates the image of time are presented by White in such a way that the emplotment by the historian, that is the subjective element, plays a principal role in the process: the historian's "narrative" gives form to historical time. Historical time does not exist outside of this narrative. The "objective" chronology of events is of subordinate importance (and indeed does not exist outside of the texts in which it is represented). "Histories gain part of their explanatory effect by their success in making stories out of mere chronicles" (White, 1985, p. 83).

White's work is thought by some to represent a turning point in the philosophy of history (Vann, 1998). "Actual experienced time" seemed to become less and less relevant to historical time consciousness. For instance, Ankersmit (1989), following in the footsteps of Mink, states that the historian should nullify actual experienced time, and he adds that well-recognized masterworks of twentieth century historiography contain very few dates. Affirmatively he cites Mink's proposition that "time is not the essence of narratives" (Ankersmit, 1989, pp. 22–23). The more a historian has to take refuge in chronological enumerations, the less successful is his work. Some narrativistic philosophers even believe that objective time is of no importance to the historian: everything is about the historian's composition, which is transcending objective time (Jenkins, 1991).

Ricoeur believes that the historian's narrative creates a bridge between the objectivating and subjectivating approaches to time: "time becomes human to the extent that it is articulated through a narrative mode, and narrative attains its full meaning when it becomes a condition of temporal existence" (Ricoeur, 1984, p. 52). Here he is referencing the analysis by Aristotle (1932) in his *Poetics* of the staging or plot (*muthos*) that is characteristic of a narrative. Although the *Poetics* is an analysis of Greek tragedy, Ricoeur believes that this applies to any form of narrative, including the historical. According to Aristotle, actions are imitated in a tragedy, but this does not imply that reality is simply copied. Actions are presented in a particular way to the viewer. This is the *muthos:* the "composition of actions" (*hè ton pragmaton sustasis*), which Aristotle (1932) believes is the most important element of a tragedy. A tragedy cannot simply start somewhere at some indefinite point in time, but it must have a well thought-out beginning, middle and end. The worst plots are those that are merely "episodic," which means that the succession of events doesn't have any logic, likelihood (*eikos*) or necessity (*anankè*) (Aristotle, 1932). (The word "episodic" should be interpreted here in its original Greek meaning: something that simply occurs and is not a part of a logical entity).

According to Ricoeur, the historian has to deal with prefigured time, that is, the passage of time as it actually occurred, but he cannot simply imitate this in his work. Time is therefore converted by the historian into configured time, that is time as it is composed by the historian in his work. The result is a reconfigured time: temporal experience as arises with the reader of a historical work (Ricoeur, 1984). The interference of the historian and his reader with the prefigured time is a necessity. After all, a historical narrative can never comprise "all of time," nor "just a time." It must have a beginning, a middle and an end, the right scope and a particular theme. It cannot simply follow events as they occurred chronologically, as that would produce a badly comprehensible entity, an entity that, according to Aristotle, would resemble a work with an "episodic" plot. History like this would resemble a medieval chronicle, in which years dominate the composition.

In contrast to the conclusions by the narrativists who strongly relativize the importance of the objective passage of time, Carr proposed the idea that events cannot be taken entirely at will as beginning, middle or end, as there are structures with a beginning, middle and end present in the actual time experienced. Experience may occasionally have the character of a "temporal chaos," when it seems to consist of no more than "one thing after the other," but "existence in time" may also have a narrative structure: "the narrative grasp of the story-teller is not a leap beyond time but a way of being in time" (Carr, 1986, p. 89). Lived time displays narrative aspects according to Carr. For instance, the inauguration of a president is clearly a "beginning" for him and for many who experience the ceremony with him, namely the start of a presidency. Our entire existence is only possible because we are constantly aware of our intentions that are associated with our image of the past and future; that's why also the really experienced time has a narrative structure. Carter (2003) argues along the same lines using an example about the purchase of land from *Aborigines* on which the city of Melbourne would later be built. This transaction was already part of a story at the point at which the purchase was made:

> Planning something or assessing what you have or haven't done involves projecting the chronology of actual or possible action into the meta-chronology of preview or review, turning the articulation of time on itself. To this extent, anticipation and memory, even protention and retention, are narrative in form even though their objects are ordered chronologies rather than meta-chronologies in their own right. (Carter, 2003, p. 15)

(Carter is referring to the shaped time with the concept of meta-chronology, which Ricoeur refers to as "configured"). According to Carter reality is indeed ready as such to be read as a story. But there are numerous stories to choose from, and these stories overlap with each other as an inextricable knot.

For Carr (1986), Ricoeur (1984), and Carter (2003) objective time and actual experienced time are relevant to a historical narrative. This standpoint appears to be plausible, as ignoring objective time could come down to the idea of "anything goes," which is clearly in conflict with the standards of historical science. The paradoxical conclusion that can be drawn from all of this is that on the one hand the experience of historical time is only completely possible through the emplotment introduced by the historian (the actual time experienced has, after all, disappeared and can no longer be experienced as such), yet on the other hand the "configurative work" of the historian cannot take place outside of "prefigured" time.

Inspired by White's (1985) essay on the historical narrative as a product of literature, the sociologist Zerubavel created an inventory of types of emplotment of time, not so much the narratives of professional historians but rather the collective memories of societies (Zerubavel, 2003). Nevertheless the types of "plotlines" he distinguishes also play a role in the work of historians and may partially be the products of the works by historians. Firstly, Zerubavel refers to "progress" and its counterweight "decline." The progress emplotment is based on the principle of "later is better," while decline is often paired with feelings of nostalgia and dreams about the good old days. The combination of progress and decline perspectives provides a composite plot consisting of a zigzag line of progress and decline, in which important "turning points" may constantly be discerned.

Linear emplotments can be contrasted with cyclical ones. In a linear development line each successive stage stands on the shoulders of its predecessor. Zerubavel comments that linear descriptions tend to become teleological and deterministic, noting in criticism that "true" historical developments are contingent. A cyclical emplotment is based on a continuously recurrent history. Although this type of emplotment of time has faded somewhat into the background in our culture, there are still numerous recurrent patterns (e.g., the recurrent four year terms of office of American presidents), and many people also have a tendency to search for analogies in developments that appear similar: history that "rhymes."

The last two types of Zerubavel's emplotments concern the intensity of the image of the past: some periods are experienced as "full" and others as relatively "empty." For instance, the collective memory of the five years of German occupation of the Netherlands would be an example of a relatively "full" period, attracting a lot of attention and many remembered facts. In association to this, "gradual" and "intermittent" developments are being distinguished. These images do not correspond to the actual passage of time according to Zerubavel, given the fact that in reality every period is equally "full."

The anthropologist Fabian accuses Zerubavel in a criticism of his work of constantly incorrectly contrasting "actual time" with the social "figments" created by societies (Fabian, 2005). According to Fabian "actual time" can never be an object of study, both because it is continuously evaporating and because the researcher himself is always situated within this time. There is therefore no basis to maneuver oneself into a position of someone who can assess what social memories do "well" or "badly," as suggested by Zerubavel.

This criticism hits the mark in the discussion about the relationship between the historical narrative and objective time. Do we only have emplotments and literary shaping, as maintained by the narrativistic philosophers? Historians find statements such as these hard to take, as is apparent from Carr's and Carter's work. According to Ricoeur (1984) they often throw "a mistrustful glance" at the "subjective time of the philosophers" (p. 227). According to Jonker (2001b), time is "more than a cultural construction, it has extra-textual aspects. If the practice of history has any hard paradigmatic core, then it is chronology" (p. 174). The "hard paradigmatic core" is a reference to the scientific character of history. However, the key role of the narrative in shaping historical time does not imply, according to Ricoeur, any sacrifice of history's scientific ambitions to subjectivism: "My thesis rests on the assertion of an indirect connection of derivation, by which historical knowledge proceeds from our narrative understanding without losing anything of its scientific ambition" (Ricoeur, 1984, p. 92). Jonker's "hard paradigmatic core" of historical practice plays a role in the configuration of time. It needs to be established how that takes place, in order to understand what could be considered as the features of the historical emplotment of time.

Ricoeur distinguishes three "reflective instruments" used by historians to create a bridge between "experienced time" and "objective time," resulting in "historical time": the calendar, successive generations, and sources and documents (Ricoeur, 1988). Each of these three can be said to have an "objective" and a "subjective" aspect: objective in the sense that they exist as a "given" and subjective as they play a particular role in perceptions of time. For instance, it may be stated that a calendar exists more or less objectively if no events are linked to it, like an empty diary a person may purchase at the start of a new year. The calendar becomes subjective once it is filled with events. The coexistence of people from different generations is an objective fact in human existence, but it acquires subjective aspects if a narrative is attached to it about "then" and "now." The same also applies to the existence of relics of the past in terms of sources and documents: the fact they exist is an objective given, but as soon as they are used for historical research they receive subjective connotations.

Three key concepts of historical time consciousness can be drawn from the development of theory about the emplotment of historical time as a mediating agent between objective time and subjective imaging:

- chronology (the calendar)
- generations, and
- relics.

Six concepts could be drawn from two principal moments in the development of historical time consciousness, namely the sense of a breach in the second half of the eighteenth century, and the focus on the role of the historical narrative in the second half of the twentieth century. They are six key concepts that may be of importance for history education. The six key concepts are depicted in Figure 2.1. The following section provides a further elaboration of these concepts in this sequence: chronology, periodization, generations, relics, anachronism and contingency. This sequence was chosen as there is a relationship between chronology and periodization, between generations and relics, and between anachronisms and contingency. The coherences between these pairs of concepts will be explored after the discussion of the separate concepts.

Six Concepts of Historical Consciousness of Time

Chronology

The calendar is a tool that mediates between the subjective experienced time and the objective, natural time. The linguist Benveniste (1974) focuses on the calendar in his description of the linguistic tools people

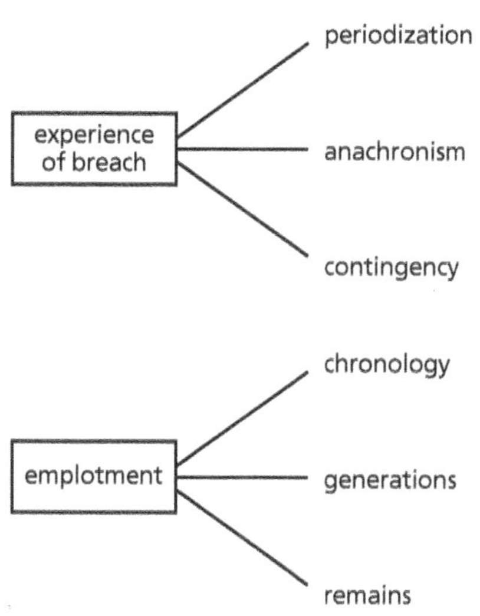

Figure 2.1 Concepts of historical consciousness of time.

apply in order to express time, as a tool to objectivize experienced time (*temps chronique*). Calendars achieve this by selecting a "zero point" from which years can be counted and oriented as before and after that point (the presumed time of the birth of Christ is taken as this point for Western calendars). They give time a direction by proposing it is a straight line to and from the zero point, and they provide tools to measure time by dividing it, for instance, into centuries. These three tools are, by themselves, "empty"—they are independent of the events they describe (in other words: all events fit within them), and, in that sense, they are objective. However, as a cultural feature they are, of course, not objective.

The choice of a zero point is tied to the sense that a new time has started. In addition to the presumed moment of Christ's birth, other points have been chosen by many cultures, including the foundation of Rome, Mohammed's expedition from Mecca to Medina, the start of the French Revolution, or the moment of world's creation as can be calculated via the traditions of the Old Testament. Other examples include the Cambodian dictator Pol Pot, who wanted to define 1975, the year in which he came to power, as the starting point, or the idea of *Berlin Stunde Null* (Berlin hour zero) after the devastating bombardments of Germany in the Second World War (Zerubavel, 2003). The use of the term "ground zero" for the area of the destroyed Twin Towers in New York perhaps rests on a similar line of thought. However subjective the choice of a zero point may be, a calendar cannot exist without a zero point.

The perception of calendar time as being linear with successive years is a cultural construction as well. Natural, cosmological time has, in principle, a cyclical rather than linear structure: the sun comes up every day and natural events have an annual recurrence. In many cultures the "eternal recurrence" of time, bound with recurrent rituals, is the dominant theme. Therefore, the cyclical pattern is more of a determinant of human temporal experience than the linear pattern of historical time consciousness (Aveni, 2002; Eliade, 1991). Important remains of the cyclical pattern are featured in Western culture still, for instance in the experience of weeks in which people "recover" at the weekend through "recreation." The weekend is in that sense "sacred" and the week "profane" (Aveni, 2002, p. 56). This also applies on a different scale for recurrent holidays and annually recurring festivals. However, for the remainder, Western temporal consciousness is shaped more by the image of a linear passage of time.

The objectivating measure that chronology imposes on temporal experiences stands at a relatively great distance to personal temporal experience. This is illustrated by the fact that we need to search to find what the date is, yet we easily remember and can locate in time the "days experienced," such as a wedding day, birth of a child or a house move. Therefore it is easier to have temporal perceptions in the form of "it was in-between the move and my new job," than in the form of "it was year x or y." The "ideational time" (time as we imagine and describe it) and "lived time" (time as we experience it) do not agree with each other, as descriptions of time more often include references to calendars and clocks than experiences (Aveni, 2002).

Chronology enables sequences to be determined but does not by itself make any distinction between past and future, simply because it keeps running. This too suggests a difference between personal temporal experience and calendar time (Ricoeur, 1988). Chronology is a B-series in McTaggart's definition, and therefore time would in fact not exist without personal temporal experiences or filling in calendar time with events. However, the objectivating measure of temporal sequences and duration also imposes limits on what kind of emplotment is possible within historical narratives. Zerubavel's criticism of what some emplotments of time accomplish does in that sense hit home. Chronology enables us to determine "where we are in time" in relation to everything else that has occurred; in addition, events that happened simultaneously may be adopted into a single entity, possibly displaying a certain level of coherence (Ricoeur, 1988). Chronology implies a disciplined way of thinking, placing boundaries on the historian's freedom to emplot time according to his/her own preferences.

Periodization

There are aspects of objective and subjective time in the activity of periodization, just like in chronological dating. The distinction between historical periods is on the one hand down to their calendar time: their starting point, their ending point, and their duration are expressed in objective time. On the other hand a period also represents something qualitative: it is characterized by certain events and phenomena, which typify it and distinguish it from others (Blaas, 2001; Ebels-Hoving, 1989). Jordanova (2006) refers to a tension between these two aspects of periodization that are inseparably linked to the historian's work. The subjective aspect is more important for periodization than it is for chronology.

The conflict between the objective and subjective features of periods becomes clear with the endless discussions about the "right" time points at which they start and end. A classic example of this is the discussion about the start and end of the Middle Ages (Ankersmit, 1984). The problem comes down to the tension previously described under "emplotments of time." To what extent was there a "temporal chaos" of only "one thing after the other" in the "actual time experienced" that can only be understood via the deliberate use of emplotments by the historian? To what extent do narrative structures exist in the actual time experienced and therefore also periodization in a rudimentary form?

These issues are expressed in the various preferences among historians concerning periodization. One could decide to opt for an as-neutral-as-possible nomenclature for time periods in an effort not to impose matters on to "historical reality" that would not do justice to it. This is what Huizinga (1950) has once argued for: each name for a time period tempting too much thought interferes with the a correct understanding of it; therefore it is best if one chooses names that "bear the seal of their groundlessness on their foreheads" (p. 93), for instance by simply numbering centuries. One could also decide, however, that any reference to the "time actually experienced" makes little sense in historical imaging, given that there is no historical time at all outside of the historian's perceptions: "historical facts are timeless and discontinuous until woven together in stories" (Lowenthal, 1985, p. 220). Such an awareness would argue for bold interpretative choices where time periods receive clear, even pregnant names, of which paradoxically enough Huizinga himself provided an example by typifying the fourteenth and fifteenth centuries as "the Autumn Tide of the Middle Ages" (*Herfsttij der Middeleeuwen*).

The problem of continuity and discontinuity in history is closely related to periodization (Romein, 1976; Ankersmit, 1984). (Dis)continuity may only be discerned by comparing two moments in time, something which can never be achieved "in the lived time itself." There is no (dis)continuity in "experienced time"; it can only be discerned on the basis of historical images. Depending on the perspective chosen by the historian, discontinuities may or may not fall under consideration. This also means that it is impossible to state that any established discontinuities enforce specific periodizations. Nevertheless, there are clearly distinct "times" in historical perceptions, which display unmistakable differences, as well as occasional agreements, all of which are dependent upon the historian's perspective. For instance, splitting up Dutch twentieth century history into a period "before the war" and a period "after the war," thereby characterizing May 10, 1940 (the date of the German attack on the Netherlands) as a turning point, is not an unavoidable option ensuing self-evidently from "the facts" (Blom, 1989). The same applies to interpreting September 11, 2001 as a turning point in world history (Jordanova, 2006).

The French historian Braudel (1979) distinguished between change processes with different tempos: evenemential (brief, dependent on incidental events), conjunctural (with cycles of one to several decades) and structural (long-term almost stationary factors). He was inclined to acknowledge a greater explanatory strength to the long-term (*longue durée*) tempo than to the superficial ripples of historical events. This is related to conceiving historical writing as an attempt to establish "how things actually were" (see under "anachronism" below). A conception of history that assumes that the past as it actually was no longer exists and has therefore become definitively inaccessible, and that consequently we can only dispose of historical imaging, should acknowledge that each of these three types of duration has an equal right to be included in a description of the past.

Zerubavel (1996, 2003) refers to "intraperiodic lumping" and "interperiodic splitting" as a phenomenon that may be associated with historical periodization. When creating an image of different periods there is a tendency to believe that the distance *between* two periods and the associated qualitative differences are large. However, anything that occurs *within* a period is often viewed as being short-lived and of the same type. Once again there is a tension between historical images and "objective time." For instance, one definitely needs an objective time calculation to become aware of the fact that the distance in time between the first Egyptian pharaoh Menes (approx. 3100 BC) and the last pharaoh, Psammetichus III (who was defeated by the Persians in 525 BC) is greater than the time distance between Psammetichus III and us. Yet we normally consider the pharaohs to be part of one category. In contrast to the passage of objective time, there is—particularly evident in ancient Egypt—the relatively slow pace of change, which makes it seem justified to typify monarchs into a single category, even if the distance in time between the first and the last of them amounts to almost 2600 years. Strangely enough, or perhaps revealingly enough, we sometimes refer to this as "the slowness of time," while the passage of objective time cannot, of course, change tempo.

Generations

The coexistence of people of different ages is a fact that can create a bridge between one's own lifetime and the lifetime of others, which started longer ago. This involves something noteworthy. Between people with whom we coexist in a single group (a family, a village community or even a nation) a feeling of fellowship, a feeling of being "us," may arise, while such a feeling cannot, in principle, exist with people in the past with whom one has not had any personal contact. Someone from an older generation is partly a contemporary and partly a historical person from another time; he is partly a person who one may know from one's own experience and partly a person who belongs to a past we cannot directly know.

The sociologist Schütz (1972) made a distinction in his writings on inter-human relationships between *Mitmenschen* (fellow human beings) and *Nebenmenschen* (neighboring human beings): respectively, people we personally know from our own experience, and people we only know through others. The latter may be further categorized into people whose pattern of thoughts we could more or less reconstruct, and those at a greater distance, for whom we have no idea what their line of thoughts looks like. The latter kind of people only exist as "ideal types"—a term Schütz borrowed from Weber. Expectations and thoughts about these people may be drawn from world experience, while abstracting from the personal; for instance, we expect a mailman to deliver the mail irrespective of his personality or his personal line of thoughts. We may have an "us-relationship" with people who are *Mitmenschen* or proximal *Nebenmenschen*, a direct or indirect us-relationship respectively, but with people at a greater distance only a "them-relationship" can exist.

It is possible to have an "us-relationship" with people belonging to an older generation, although they have partly lived in a time preceding our own. They can tell us about the people they personally knew, so an indirect "us-relationship" can originate with them. However, ultimately we pass into the domain of people in the past with whom only a "them-relationship" is possible. Schütz makes the distinction between the world of contemporaries, successors and ancestors. The world of ancestors he defines as that which existed before someone was born. The separation between the world of ancestors and those of contemporaries is not absolute, as older people from a "different" world are alive in the contemporary world.

The world of coexisting generations is more complicated than is apparent from the distinction between contemporaries and ancestors. The sociologist Mannheim (1952) refers in this respect to the work of the historian Dilthey and the art historian Pinder.[19] Dilthey emphasized that people from different generations in a single time period still enjoy the same experiences and surroundings through one and the same *Zeitgeist*, while Pinder (1961) advanced the idea of the "non-contemporaneity of the contemporaneous": each generation lives in its own personal life phase with its own internal goals (*entelecheia*), that differ from those of other generations even if they are contemporaries (pp. 57–58). Carr (1986) argues along the same lines when he points out that the coexistence of generations boils down to people "with different overlapping narratives" (p. 113) living at the same time, each at a different stage of life with their own backgrounds and aims. Giesen (2004) refers in this sense to *divided memories*. In his view older generations should be taken as ancestors more than as contemporaries, because their memories are essentially different from those of younger people. Nevertheless "us-relationships" may exist between different generations. According to Carr (1986) older people, in turn, have had similar relationships with their ancestors, giving rise to a kind of historical continuity where inheritances are constantly passed on, paired with an "us-relationship" in the long-term. Something similar is stated by Zerubavel (2003).

The question is how the issue of "us-relationships" and "them-relationships" can be connected with historical time consciousness. Carr refers to what he calls a "pre-thematic historical awareness," an awareness of the past that has not resulted from deliberate study, which he nevertheless refers to as "historical." This contrasts, for instance, with Oakeshott (1983), who only refers to the past as "historical" if it arises from deliberate study. This may imply that Carr conceives of "history" in a broader sense than it has been defined in this book, namely academic history, which is the result of deliberate study. Schütz intended to describe how the "us-" and "them-relationships" function in human society at a particular moment, which, among other things, may be viewed as a society of ancestors, contemporaries and successors. He was not interested in describing and analyzing a consciously studied historical past. All of this may provide a basis to put Schütz's and Carr's positions regarding "ancestors" that allow "us relationships" into perspective. It has been argued above that a sense of a break with and distance from the past belongs to historical time consciousness; the development of an "us-relationship" even with people we do not personally know heads in the opposite direction. It may therefore be presumed that a historical time consciousness is based on the distance of a "them-relationship" rather than the proximity of an "us-relationship."

How meta-historical is a group? Jordanova (2006) describes how, particularly in the United States, the view is sometimes taken that Jewish history should only be written by Jews, African-American history by African-Americans, and women's history by women, as an "outsider" would not be able to understand these histories. However, this would imply the assumption that women, African-Americans or Jews are timeless categories. This brings us to a similar paradox as the one discussed earlier in the context of hermeneutics. History involves the description of "other times" wherein the "them-relationship," rather than the "us-relationship," is characteristic of the relation between the historian and his/her subject. There is therefore no reason that women's history should be written by women, African-American history by African-Americans and Jewish history by Jews.

The statement by Russell (2000) in his *Autobiography* from 1967 that he could still remember conversations about his grandfather visiting Napoleon on Elba from his earliest childhood evokes a strange sensation. After all, Napoleon belongs to the world of history, and conversations with grandfathers belong to the social world, which are normally viewed as being distinct. The relationship with a historical person is not, in principle, an "us-relationship" as occurs in the social world; rather it is a "them-relationship" as exists with people we do not personally know, or in the case of history, people we *cannot* know in person.

The ultimate consequence of a historical time consciousness is to see even oneself in the past as "another person" with a (partially) different identity: "this type of transition therefore confronts us with the confusing situation where it appears as if someone has suddenly become someone else—and he views his former self as if it were someone else" (Ankersmit, 2007, p. 380). The Dutch novelist Nooteboom gave an example of this in his acceptance speech to the Adenauer Prize on December 12, 2010 in Weimar. He noted in a reflection about himself and the German Chancellor:

> For a few years an old German politician and a young Dutch author were contemporaries. I never realized it at the time, nor did he, but now, perhaps due to the prize, perhaps because the name is more in circulation recently, he's keeping me busy. (Nooteboom, 2010, n.p.)

It is an example of historical thought in terms of contemporaries and ancestors, of being able to see oneself in the past as a different person that once was, and that can be put at a distance.

In contrast to Schütz and Carr, my inclination is to make a distinction between people with whom we have an "us-relationship" as belonging to the world of contemporaries, and people with whom there is a "them-relationship" as belonging to the world of ancestors. This may in certain instances also be one's own previous world, if one has the sense that in a previous stage this was essentially different to the present world. The "them-relationship" would then be characteristic of a historical time consciousness, even if that implies a "them-relationship" with one's own previous self. Ankersmit (2007) has provided a sound description of what the border between the present world and an earlier one is—in other words, what is the "extent of the present," using the image of a stretched elastic:

> It may happen that we suddenly and surprisingly conclude that at a particular moment in human existence the past has dissociated itself from the present without us noticing it... At a certain... domain of human existence the present has been stretched and stretched further, until the point has been reached when the elastic between the past and the present snaps and divides itself into one part, the past, and another, the present. (p. 420)

If historical time consciousness is defined in association with the "them-relationship," then this counteracts people's tendency to extend the group to which they belong back to the past, without limits, for instance, by referring to the sixteenth century Dutch Revolt as if it were a conflict between "us" (the Dutch) and the Spaniards or even viewing the Batavian Uprising against the Romans in the first century as an example of the "civil unrest" characteristic for "us, the Dutch." The extent of the temporal distance is not a decisive factor here. A historical approach to the occupation of the Netherlands during the Second World War would imply not to see it as a confrontation between "us," the Dutch, and "them," the Germans (both as a meta-historical concept), but between former Dutch and former Germans, even in the case where views of the occupation are provided by someone still alive who has lived through that time.

Relics

The world of experiential time is not only filled with human ancestors and successors, it is also full of relics from other times. Geographical features, such as the course of rivers, the presence of seas and harbors, hills, and so on are the first category belonging to this; these have changed very little over the course of the centuries and as such create a natural connection with other times (Zerubavel, 2003). It is no coincidence that Braudel based his argument about the long-term, virtually unchanging factors in history on a study of geographical features of the world of the Mediterranean Sea (Braudel, 1966). In addition to the geographical situation, there are also cultural relics from other times, such as monumental buildings and ruins, traces in the landscape, objects of art, tools, and so on. These relics may function as a bridge to other times if they are perceived and recognized as such (Lowenthal, 1985). In actual fact the entire world around us may be perceived as a relic from other times, but usually the specific study of it in this sense is reserved for certain historically defined objects (Oakeshott, 1983).

This applies in particular for documents stored in archives and viewed by historians as "sources." These documents create a bridge to other times because, though they exist in the present, it may be *considered*—in that same present—that they were created in the past as an expression of the authors' particular thought processes. Collingwood used this to distinguish between the real existence (in the present) of historical relics and documents and the imagined existence of them in the past; these are the "realistic" and "idealistic" dimensions of the existence of historical evidence (Van der Dussen, 1988).

The imagined existence in the past of relics is crucial. The present day existence of the material as "historical" is not essential; rather it is the way in which it is questioned by the historian (Collingwood, 1973). The latter transforms it into historical evidence. In principle, any item in our world may be considered for use in this way. The distinction made by Oakeshott between a "practical" and "historical" past is heading in the same direction.

The practical past is present permanently and unconsciously and determines expectations and actions in roughly the same way as St. Augustine's threefold present and Husserl's retention and protention included in the "now." We may refer to a historical past if the present is *intentionally* being studied with a view to developing an image of the past, as is the case for Husserl's "secondary memories" (Oakeshott, 1983).

The inclination to develop a faithful image of the past through applied research rests on a particular attitude that is not easily developed in everyone, according to Oakeshott. Carr (1986) is also of the opinion that the usual "social past," even if the image of this is intentionally evoked by placing activities and perceptions from the present within its context, has to be distinguished from a conscious thematic study of the past for its own sake.[20] According to Oakeshott (1983) this type of study implies the "most sophisticated" attitude one can adopt, "difficult to achieve" and "difficult to sustain," and also highly susceptible to relapse "into some other engagement" (p. 28). The latter is intended to mean dealing with material from the past without really taking the temporal dimension into account, which is a tendency that features dominantly in most people.

Anachronism

Huizinga (1948b) once stated that the avoidance of anachronisms represents one half of historical science. This characterization of the historian's work is connected with the requirement, such as that formulated by the nineteenth century historicist, to do justice to each individual period in itself. The most famous and most often cited observation in this respect is that by Ranke in his foreword to the *History of the Romance and Germanic Peoples Around 1500.* According to Ranke (1885), history had often been expected to pass judgment on the past or to make it possible to draw lessons from the past for the present and future: "this work does not attempt to undertake something as weighty as this, it only wants to show how things actually were" (p. VII, my transl.).[21] Historians wanted to efface themselves entirely and let the past speak for itself, while carefully avoiding any "bias" being introduced by perspectives from their own times.[22] Nowadays we believe that such a thing is impossible. The only view of the past than can exist is a view from a later perspective. "History itself" is unable to speak, as an earlier reality does not exist (anymore).

One *has* to employ "anachronisms," categories and concepts that do not belong in a particular past, in order to help modern day people understand other times. This has to be the case, if only because of the modern day language in which events are described (Blaas, 1988). Nevertheless the requirement to do justice to each time period "as it really was" is still valid in historical sciences; it is something that historians pursue, even though they are aware of the fact that it can never be achieved. Most historians reject the relativistic idea of narrativist philosophy of history that states we only have literarily shaped stories and that there is no essential difference between history and fiction (Vann, 1998; Vries, 1990).

The problem of anachronism plays a particular role as far as values and judgments are concerned (Vermeulen, 1978). Weber's distinction between *Wertungen* and *Wertbeziehungen*—value judgments and (description of) value relations—is a classic in this respect (Ankersmit, 1984; Van der Dussen, 1988). History always involves dealing with value-laden issues. However, the historian has to confine him/herself to describe the role of values in the past and should not to be tempted to given his/her own value judgment. Although a value-free historical science is difficult to imagine, historians strive to avoid moral judgments, motivated by the conviction that one time period does not have the right to judge another, as each time period has to be understood by itself. A historicist standpoint implies that meta-historical values do not exist. History should, some people believe, in the first place aim at understanding and explaining, not at passing judgments (Blom, 1986). That is why it is so difficult to historicize a subject such as the mass murder of the Jews during the Second World War, and why lessons on that topic seem to be an exceptional category compared to "ordinary" history lessons (Blom, 2010; LaCapra, 1988; Totten, Bartrop & Jacobs, 2004). Other topics, such as the transatlantic slave trade, are also strongly tied to the debate about values and the education of students (Wilschut, 2005).

The avoidance of anachronisms and the struggle to understand each time period by itself would be an argument for a relativizing, historicizing standpoint, striving towards freedom from value judgments. The question of how it can be explained that the murder of millions of Jews or the transport of millions of slaves could take place is a historical question, rather than the question of how we can avoid something like that ever happening again or the question of how terrible these events were and what moral lessons we can draw from them.

Contingency

Since Butterfield's publication in 1931 of the essay of the same name, the *Whig Interpretation of History* has become a permanent concept among historians (Butterfield, 1973). It represents a perception of the past as a prehistory of the present, in such a way that clear developmental lines may be drawn from the past towards the present. Anything that has "already" been the case in the past is included approvingly into the developmental line; anything that resisted certain future-referencing developments is viewed as an obstruction against unavoidable progress. Long causal chains are sketched where—for example, in Britain—the sealing of the Magna Carta is viewed as the first step on the way to a parliamentary democracy as currently exists, instead of explaining the charter as a document from feudal times with clear features of that period.

In his essay Butterfield (1973) devotes a great deal of attention to the Protestant Reformation of the sixteenth century, which according to the *Whig Interpretation* is seen as something like the starting point for the entire modern Europe: the capitalistic moral, the separation of the church and state and secularization—all these issues arose from the Reformation. However, this explanation does not do any justice to Luther, who, according to Butterfield, should first and foremost be seen as a sixteenth century religious leader and who would be extremely shocked about the state of the world today allegedly created through his intervention, as indeed would the Catholics of his time.

According to Butterfield, the course of history is infinitely complex and full of incomprehensible winding roads, backdoors, coincidences and unintended consequences. The unpredictability of the consequences of human actions is perhaps one of *the* most important lessons of history. Ignoring this and summarizing history as a straight line of progress towards the future creates circular reasoning: one starts from the present and ends up back in the present. This leads to a missed opportunity to discover deeper insights into oneself as a human being, in other words, that humans are relative and a product of historical coincidence, which could have taken a completely different course.

Huizinga had already employed that insight in relation to the continuous line assumed to exist between the Burgundian Netherlands and the later independent states of the Netherlands and Belgium. According to Huizinga (1948a) that historical development was by no means destined:

> Why not the Burgundian empire, as dreamed of by Charles the Bold, why not a nation whose language would have been transformed by the French, as the Norman French had transformed the language in England some centuries before? Only the sheer vicissitudes, the death of princes, their talents and their passions, and a thousand other circumstances have destined the further growth of the Netherlands. (p. 98)

He repeated this statement more emphatically in a later French contribution on the same topic: one should avoid all forms of determinism in history and realize that events could have taken a different turn at any time (Huizinga, 1948b).

The fact that developments take a capricious course is connected to a great extent with the unintended consequences of human actions. According to Ankersmit (2007) there is a "close relationship" between the rise of historical consciousness and the rise of insights into unintended consequences. The observation that there is a discrepancy between intentions and outcomes is, after all, always connected with a temporal difference: someone's intentions at a given time were focused on a specific result and only at a later stage it becomes apparent that the result has not been realized. A comparison of both perspectives (and thereby viewing someone at one point in time as being different than at another point), creates an insight into "unintended consequences."

Butterfields's observations can be considered alongside the demand by historicism to view each time period by itself. There is a coherence between historical time consciousness and the awareness that lines between the past and the present should not be drawn in a too direct and too simplistic way. For instance, according to Butterfield (1973) the question should not be "where does our religious freedom come from?" but rather, "why were there religious persecutions in the sixteenth century?" The first question is about the present, the second is a true historical question (although it should be noted that a phrase like "religious persecutions" also betrays a modern day perspective). However, explaining the motives of the religious persecutors also seems to make them more acceptable in one way or another: "It is in this sense that he [the historian] is always forgiving sins by the mere fact that he is finding out why they happened" (Butterfield, 1973, p. 22). This may sometimes lead to tricky problems. For instance, if one takes the effort to explain why Hitler pursued the Jews, this may lead to seemingly "forgiving" his deeds, because they are explained and being made understandable. For similar reasons Blom was accused of "malevolence" because he attempted to write the history of the Netherlands in the Second World War without employing the usual moral judgments in terms of "good" and "bad" for, respectively, resistance and collaboration (Blom, 1986).

An awareness of the unpredictability of the course of historical processes and unintended consequences of human actions seems to be an important lesson of history which may help people in their orientation on the present and plans for the future. The historical coincidence of people, their situations, their perceptions and values may play a beneficial role, introducing a sense of perspective in the discussion between social groups in a pluriform society.

Conclusion

Historical time consciousness may be defined by using the experience of a breach that led to the rise of Western historical consciousness and the way in which historians create an image of the past that is subjective in nature on the one hand, yet on the other is partly determined by the "objective" boundaries of chronology. From the experience of breach and historical emplotment, six key concepts can be derived which further define the consciousness of historical time. The sense of a breach with the past led to the distinction of anachronisms, the sequencing of different periods with their own features, and a sense of the unpredictability and capriciousness of historical development. A study of the way in which historians emplot time led to conclusions about the role of chronology, dealing with generations (resulting in a "them-relationship" with historical persons) and a perception of the world around us as consisting of relics from the past. The six concepts used here to describe historical time consciousness relate to *the picturing and division* of time with the use of a timescale and the distinction between various compartments (chronology and periodization), the *mediating elements* enabling images to be created from one time period from the perspective of another (generations and relics) and the *interrelationships* between the different individual time compartments (anachronism and contingency). This is why the six key concepts were discussed in the order described above. The first two concepts deal with the way in which we design images of the past on a linear timescale and differentiate these according to the features of diverse compartments of time; the second pair deals with the means we employ to create a bridge between various time periods, particularly the present and periods in the past; and the final pair deals with the attitudes and perceptions affecting the interrelationships between the compartments. Figure 2.2 demonstrates how historical time consciousness is summarized in line with the six concepts discussed.

The present section has presented a number of comments about the patterns of thought that logically flow from a historical consciousness of time and modes of thought that conflict with it, for each of the six concepts. If learning a historical time consciousness is a goal of history education, then this distinction will have consequences for the aims of that teaching. In the next section historical time consciousness will firstly be compared with other potential temporal experiences in order to further clarify the specific nature of historical time consciousness. Subsequently the six concepts of historical time consciousness will be applied to history education.

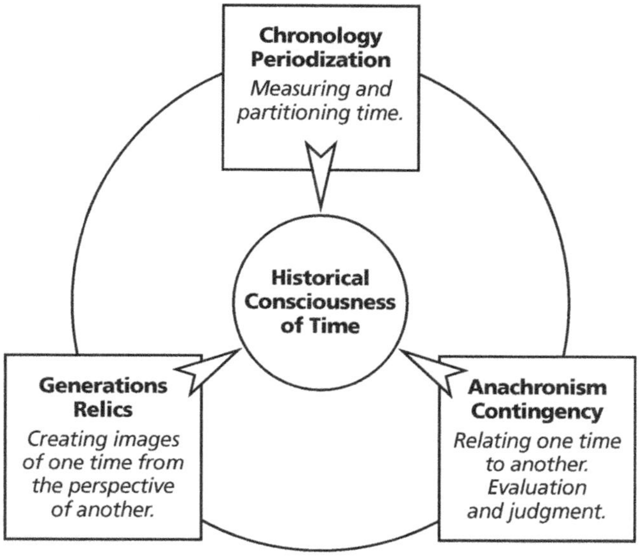

Figure 2.2

Alternatives to a Historical Consciousness of Time

The historical time consciousness as described in the previous section is usually defined as Western. For instance, in an introductory article to an "intercultural debate" about Western historical thinking, the historian Burke (2002) referred to the linear image of a series of consecutive times each with their own characteristics, the underlying developmental perspectives, the sense of anachronism and the focus on the differences between individual cases, as examples of Western perceptions of time. Some even believe that there is a Western "prejudice" in this, whereby "the presumed superiority of modern, industrial society can be retroactively substantiated" (White, 1973, p. 2). The anthropologist Fabian (2002) has suggested that this Western notion has denied other cultures the right to be "contemporaries" (in the sense of belonging to the same environment, shaped by time, as Western culture). According to classical, predominantly nineteenth century anthropology, other cultures were supposed to be "still" living in a previous stage of development that had "already" been experienced in the West, so that a trip through space to a different culture would also imply a trip through time. Later, anthropologists would start to study cultures as "worlds by themselves," with their own time systems not related to any universal time system (Fabian, 2002, p. 39). This circumvented the problem of contemporaneousness. It is not the intention here to further elaborate this theoretical problem or to solve it. The discussion that follows of a number of anthropological studies, presumably of the type that, according to Fabian, tries to circumvent the problem of contemporaneousness, has only been included here to review which other forms of temporal experience, in addition to a historical consciousness of time, are possible.

Research into non-Western cultures has highlighted that time, defined as a homogeneous, continuous, objective, "Newtonian" phenomenon, is much less natural and obvious than Western researchers have often assumed. Time is a cultural construction: different cultures deal in different ways with time, and the Western approach, which assumes that time is a homogeneous, continuous, linear and exactly measurable phenomenon, is one of these. One question within anthropological research is then, to what extent there is something common in the way time is experienced by humans. Initially researchers believed that non-Western cultures did "not have an awareness of time," as they did not depart from something measurable and objective (Evans-Pritchard, 1939, p. 208). More recent studies have, however, argued against this: that there exists something common in the experience of time is demonstrated by the fact that it is possible to communicate with other cultures about time. "No awareness of time" is therefore a Western prejudice, which actually means: their awareness of time is different to ours (Howe, 1981; Perkins, 1998).

What types of consciousness of time did anthropologists encounter among non-Western cultures? A classic study in this field was that undertaken by Evans-Pritchard (1939) among the Nuer, a herdsman tribe in Southeast Sudan and Western Ethiopia. He distinguished two principal categories in the way the Nuer dealt with time: an "ecological" time and a "structural" time. The first concerned the way in which economic activities were ordered during the year in line with the change of seasons: a dry time and a rainy time, which entailed different activities. The Nuer made a distinction between months, but the fact that a lunar year does not correspond to a solar year did not create any problems. If a particular seasonal activity was required, then it had to be "the month for that activity": "the activities regulate the calendar instead of the calendar regulating them" (Evans-Pritchard, 1939, p. 201).

Ecological time had a cyclical structure as it concerned annually recurring events. This differed to structural time, which was based on the coexistence of people from different generations. Although the Nuer did not know precisely how old someone was, they did distinguish between different stages in life: young children up to puberty, puberty up to marriage age, adulthood, and old age. These groups were not characterized by the exact number of years someone had lived; rather they were distinguished by their different status in society. Each age group had its own unique name, which remained with the group as long as they were alive. Such a group would roughly span twelve years, and members from no more than seven groups would be alive at the same time, creating a total time span of around ninety years. The names of groups of which people were no longer alive were quickly forgotten.

Individual events (the birth of a calf, a cow attacked by a lion) were easily recalled, but it required a great deal of effort to calculate precisely how many years ago they had occurred, as there was no counting of years. This brought Evans-Pritchard (1939) to the conclusion that a consciousness of historical time did not really exist among the Nuer: "Historical time, in this sense of an order of events, probably never goes back among the Nuer more than fifty years" (p. 210). Anything referring further back in time belonged to the category of the "far, distant past." This was a mythical past, which could not be truly placed in time, as the distance between that past and the present was always perceived as being the same. It was the time when the world was shaped by the gods.

The findings by Evans-Pritchard have been confirmed by other researchers. Hallowell (1937) studied a culture in a completely different region than that of the Nuer: the Saulteaux Indians along the Berens River in Canada, to the east of Lake Winnepeg. Their temporal awareness was also primarily based on seasonal activities, loosely associated with a division of the year into months, where the activities or seasonal events—similar to the situation among the Nuer—were decisive in determining the month (the "month in which the geese return") and not vice versa. This was also used to solve the issue of the differences between a solar and a lunar year. The Saulteaux separated people into age classes, without knowing their exact age. They did recall information about their forebears, but the time span traversed in this was limited. Hallowell (1937) estimated a greater length of time for the period about which knowledge was retained in this way than Evans-Pritchard: "... information of this sort completely fades out in less than two centuries. One hundred and fifty years is the outside limit of any historic past so far as the Berens River Indians are concerned" (p. 667). Anything that fell outside of this belonged to the category "distant past," a mythological time without any chronological significance, involving godlike beings and demons who existed then as well as now, thus transcending time.

Research from the 1970s provided results that were in line with those of Evans-Pritchard and Hallowell. For instance, Turton and Ruggles (1978) investigated temporal awareness among the Mursi, a tribe of farmers and cattle herders in Southwest Ethiopia. They focused on the way in which the Mursi solved the problem of the difference between the lunar and solar year. The Mursi tribe counted the months and also linked particular seasonal activities to certain months. However, their counting led to the conclusion that a year is sometimes twelve months and sometimes thirteen months. The solution was to leave the answer to the question which precise month it was vague and to allow any differences about this to "experts," who might have different perspectives among themselves about the exact month. Similar to the Nuer, the activities in the context of the changing seasons were the decisive factor. The Mursi also distinguished between age groups in their social life, and age was defined by the group someone belonged to, rather than by a precise calculation of the number of years of someone's life.

Around 1970 Ohnuki-Tierney (1973) studied the Ainu, who originated from the Sakhalin peninsula, but had been brought across to Hokkaido in Japan. The Ainu made a distinction between a "distant past," a mythical "start of the world," and a "present" which were separated by a gap that could not be defined in terms of time. The mythical "start" was a "holy period" that had little to do with what "Westerners call time, history or chronology" (Ohnuki-Tierney, 1973, p. 287). Seasonal economic activities and events played a key role in the present. Similarly to the cultures discussed earlier, activities were linked to months, but the Ainu were not precisely aware which mathematically defined month was being referred to. In comparison with the Nuer and Mursi, the structural temporal awareness of the Ainu was only moderately developed, although they did distinguish between different age groups. Memories of events appeared to be poor: "the Ainu concept of time is basically qualitative; quantitative measurement of time is little developed" (Ohnuki-Tierney, 1973, p. 292). There existed no word for "time" in their language (nor was there, indeed, in the Nuer and Mursi languages), and their verbs did not have different tenses; however there was a very extensive system of modalities in the verbs to characterize the nature of actions: short-term, recurrent, successive, concurrent, and so on.

A remarkably different image of time than that usually employed in Western culture was shown in a study of the Aymara Indians, who live in the Andes mountain range in the north of Chile, southern Peru and southwest Bolivia (Nuñez & Sweetser, 2006). In Western languages time is usually defined by means of spatial metaphors, in which either the speaker moves forward along a line (as in "we are approaching the end of the year"), or the speaker is standing still and time is moving along (as in "the end of the year is drawing close"). In both instances the future is seen as something that lies ahead of the speaker. This appeared to be completely opposite for the Aymara. They depart from a static speaker as a reference point, with the past ahead of him and the future behind. The latter became apparent among other things from their sign language: when speaking about the future the Aymara pointed backwards, and they pointed forward when talking about the past.

There was no evidence in the non-Western cultures described above of an abstract concept of time as something that exists by itself and occurs irrespective of the events that take place within it. In these cultures the "calendar" does not show when something is happening or should be happening; it is the events themselves that make clear what time it is. "Empty time" does not exist in this conception. There are also non-Western cultures where an abstract conception of time does plays a role. For instance, traditional Balinese culture calculates time in such a way that every day is part of ten different time units of unequal length, in the same way that each day in Western chronology is part of three overlapping units of time: the week, the month and the year. The Balinese calendar system determines at which time certain events have to take place, for instance, the Galungan ceremony that takes place when the five day, six day and seven day cycles coincide, which happens every 210 days (Howe, 1981). Similar to the other cultures described above, Balinese temporal awareness places an emphasis on the continuously recurring, cyclical element of time.

The Mayan culture, another non-Western culture with an advanced calendar system, also had a number of overlapping cycles (Aveni, 2002). Around the second century AD a "zero point" was calculated for this system: a point at which the world was created—in the same way that the Old Testament Jewish scholars determined the moment of creation (Aveni, 2002).[23] From that point onward there was an interplay of cyclical time units, which continuously added up to "large cycles" of 144,000 days (approximately 394½ years). However, this did not imply any vision of the present, past and future, as in Western culture (Aveni, 2002).[24] The Chinese calendar system demonstrated a combination of cyclical elements with a long, chronological line of rulers and their deeds, carefully noted down by the emperor's bureaucracy, but without creating a historical consciousness that is comparable to the one discussed in Section 3 of this chapter (Aveni, 2002; Loewe, 1999).

Conclusion

Temporal awareness often appears to consist of a cyclical element determining recurring events and activities—particularly under the influence of an annual cycle with changing seasons—and a linear element that arises from the fact that people of different ages live together. The fact that different age groups have a different status within a given society is particularly important in this context, the memories passed on through the generations being only secondary to this. The awareness of time that arises from different generations living together may be typified as "social" time and the awareness of cyclical events as "daily time" (structural and ecological time respectively in Evans-Pritchard's definitions). The length of recall associated with social time has been estimated by anthropologists to be no greater than roughly 150 years. This follows from the ages reached by humans. Outside of both temporal categories, there is an awareness of a mythical, primeval time, which can only be defined as having occurred "long ago," without precisely indicating how long. The dating of this time is felt as irrelevant, because it involves a mythical time and not history or chronology as usually perceived in Western societies. It concerns "long, long ago," but there is also a sense of eternity about it. The latter is emphasized by the ritual re-enactment of the act of creation (Eliade, 1991).

The three types of temporal awareness, the recurring daily/annual cycles, the social awareness of memories of those with whom we share our lives, and the mythical awareness of a primeval era from the "distant past" do not concord with historical time as described above. Social time is a borderline case here. It is linear and partly determined by chronology as is the case for historical time. Studies of non-Western cultures did reveal they were able, in certain situations, to calculate how long ago something had occurred or how old someone was by pulling together all kinds of fragmentary details (Evans-Pritchard, 1939; Turton & Ruggles, 1978). However, in contrast to a historical consciousness of time, these cases did not involve a period of time vis-à-vis which a sense of breach was felt or a period of time of which the people living in it were considered as a them-group.

The distinction between Western and non-Western temporal experiences is relative in the sense that it cannot be stated that every form of Western temporal consciousness is always fully "historical." Cyclical temporal experiences—such as the cycle of weeks and years—as well as a social temporal awareness occur in abundance in Western societies, and also perhaps even the "mythical" experience of a "primeval time" in cases where there are vague and undifferentiated images about "the past," and this "past" is contrasted with "now" as a more or less fixed entity. A consciousness of historical time might be a strange, exceptional category, an artificial way of thinking about time. The contrast between a historical consciousness of time, as defined in six key concepts, and alternative temporal experiences described above is shown in Figure 2.3.

In opposition to the awareness of recurrent, cyclical time of, for instance, weekly or annual events, the historical consciousness of time sug-

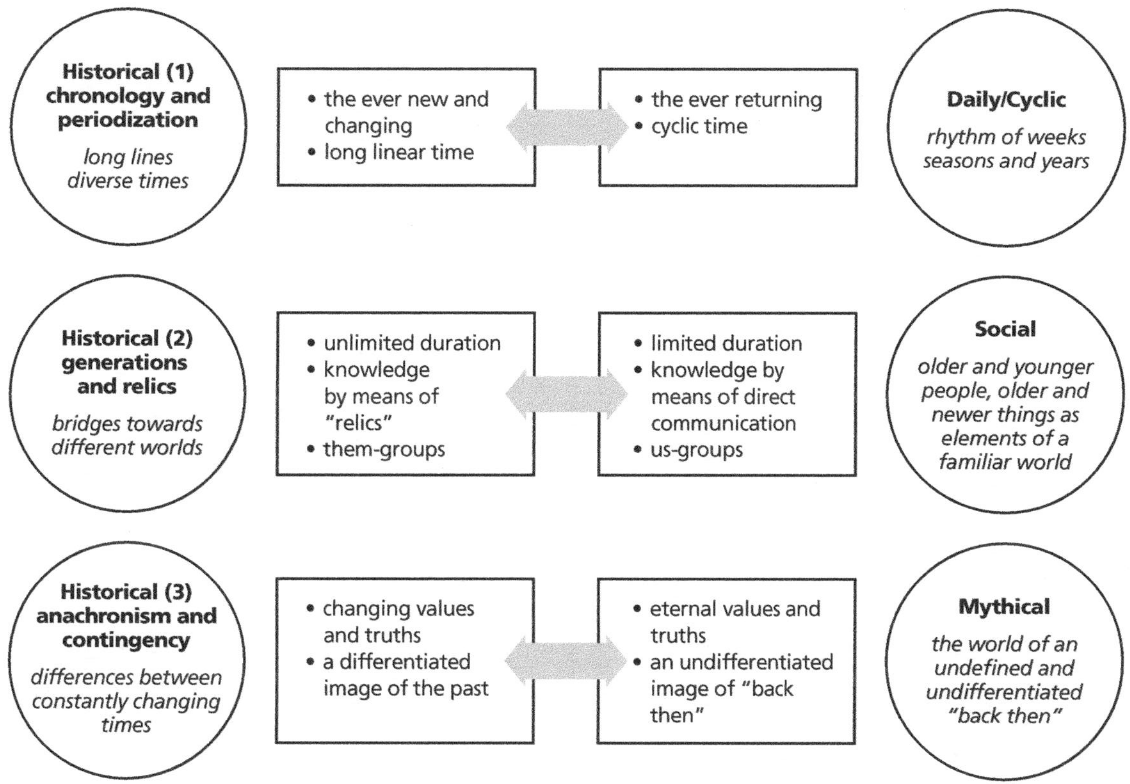

Figure 2.3 Consciousness of historical time and its alternatives.

gests the idea of constantly different, changing and novel events and phenomena that are placed on a long, in principle endless line.

The historical image of generations and relics contrasts with the awareness of social time based on "us-group" relationships and well-known objects from a familiar world. Even though people of different ages live in the world of the present, and though older and newer objects co-exist in it, they all principally represent a familiar environment, which does not invoke any sense of distance. In contrast to this, the historical consciousness of time is based on the idea of the other and of others: other people with whom we have a "them-relationship," and strange relics and documents creating a bridge to other worlds that are essentially different from the present. An effort is needed to view older generations and relics from other times as "alien," for instance to imagine that grandma once was a young girl, or to picture how an old church has been brand new hundreds of years ago.

The awareness of mythical time departs from a world that was "once," an indeterminate past often linked to fixed convictions and centuries-old truths. A fairy tale, for instance, starts and ends at certain indefinite points in time ("Once upon a time . . . They lived happily ever after"), but conveys messages about good and evil and human nature that are valid for all times. A historical consciousness of time, on the contrary, focuses on the changeability of values and truths and points to the capricious course of historical developments, where different situations and unexpected processes continue to follow each other with limited predictability.

This comparison between a historical consciousness of time and its alternatives may be important from an educational perspective. The experience of historical time may be contrasted against a number of features of other temporal experiences from the students' world outside of history lessons that students probably bring along: a world that is determined by continuously recurring weekly and annual cycles, by stories about the past from the older generations, and by vague notions about then and now, sometimes resulting in adherence to centuries-old, established truths, the simplest of which could be something like "things used to be a lot worse in than they are now, people in the past usually were poor and repressed." If learning to develop a historical consciousness of time is an aim of teaching and learning history, then that should imply that learning processes will need to focus on replacing or adapting images from the right-side of Figure 2.3 with images from the left. However, the validity of the representation in Figure 2.3 as a framework for teaching and learning history is a hypothesis in the sense that not enough is known as yet about the extent to which and the way in which images of the right side of Figure 2.3 play a role with students. Further research is needed to determine whether this hypothesis holds.

Historical Time Consciousness in History Education: Aims and Research Questions

If objectives for history teaching are derived from the previous description of the historical consciousness of time, then this description is ascribed a normative role. It is important to reflect on this for a moment before any further investigation of the role of historical time consciousness in history teaching is undertaken. Can expertise in teaching and learning history result into the formulating of goals for this teaching and learning? Should not the formulating of goals be left to others, while the expert of teaching and learning history concentrates on developing the best teaching methods based on empirical research?

The American project *How People Learn,* which was referred to in the introduction of this book, was aimed at collating the results of empirical educational research from the last decade of the twentieth century and making this applicable in practice. The project came to an interesting conclusion: "Asking which teaching technique is best is analogous to asking which tool is best—a hammer, a screwdriver, a knife, or pliers" (Donovan, Bransford & Pellegrino, 1999, p. 18). Just like with tools, the question for education is what it is intended for. The aims of teaching need to be included into the considerations, particularly when educational research aimed at domain-specific ways of thinking and reasoning is involved.

The extent to which the objectives of history education play a role in research into expertise in teaching and learning history is apparent from the following example. Jacott, López-Manjón and Carretero (1998) studied the way in which students deal with causality. In this research they distinguished between an "intentional" and a "structural" explanatory model, based respectively on people's motives, intentions and beliefs, and on economic, demographic, social, religious and political conditions. They based this distinction on history theory literature. Their empirical research demonstrated that people who have not studied history and younger students tend to mention intentional causes in historical analyses, while history students show a more significant tendency to invoke structural causes. The latter led to advice about the way in which history teaching could help promote the analysis of structural causes. History students and other "experts" dealt differently with causality than the younger students because they had *learned* to do so when studying the subject. The professional rules had therefore had a normative effect on the aims being striven for in teaching. This normative influence has also been demonstrated in other studies into expertise in teaching and learning history, in particular those where "experts" are compared with "novices" (e.g., Wineburg, 1991, 1998; Perfetti et al., 1995; Van Boxtel & Van Drie, 2004).

As discussed in Chapter One, the norms governing what has to be achieved in history teaching are determined by politics and society, which need to decide whether to respect history as an independent professional discipline of its own, or whether certain ideological goals have to be imposed on history education even if this results in ignoring the rules of the discipline. If one assumes that lines of thought belonging to a historical consciousness of time are characteristic of history as an academic subject, and if it is the aim of politics and society that history is taught at schools, then this implies that these modes of thought ought to play a role in the aims of history teaching. The next section details what shape these aims could take, including the associated questions in terms of research into teaching and learning history.

Educational aims could on the one hand be drawn from certain modes of thought and reasoning specific to a discipline, and on the other from what has been learned from research about students' misconceptions and learning problems. A combination of these two may result in adjusting students' modes of thought and reasoning in a desired direction. For these reasons the discussion below will present educational objectives that on the one hand have been based on the typology of the six concepts of historical time consciousness as described in Section 3 of this chapter, and on the other on details from research into teaching and learning history. Chapter 3 will discuss the results of educational research that can be associated with the six concepts of historical time consciousness in more detail.

The following section will discuss the modes of thought and reasoning that flow from the above considerations. This does not, however, imply right away that these are desired educational objectives; that would depend, among other things, on empirical research with students and teachers. Students would need to be studied, as this would demonstrate to what extent they already display certain behaviors (so that no teaching needs to be provided in that area) or to what extent that behavior is so difficult to achieve that it is not realistic to define it as an educational objective. Research with teachers should be able to demonstrate to what extent their modes of thought coincide with a historical consciousness of time. If it turns out that this is not overwhelmingly the case, then it could be concluded that experts do not display certain behaviors, which either would imply that it should also not be an educational objective or that teachers need to have further training.

Chronology

As explained in Section 3, chronology is based on the subjective choice of a starting point in time, yet it allows for historical sequences, (approximate) coincidences of events in time and the duration of time periods to be determined in a relatively objective way. These features of chronology lead to a disciplined way of thinking that is one of the features that need to be learned in history lessons. It has been argued above that people more easily remember dates they have experienced (the day I moved house) than dates presented on a calendar (April 11, 1990). This distance between calendar time and experienced time, the "artificiality" of chronological dates in relation to the "naturalness" of "dates experienced" is perhaps a starting point for research into the way in which students learn to deal with chronology. Furthermore, it was noted above that there are numerous cyclical elements in the awareness of daily time (the rhythm of the week, monthly salary, annual recurrence of seasons and holidays) and that a long, linear passage of time is only reflected upon at specific occasions. This also represents a starting point for research into learning chronology. There has already been quite a bit of research into the way in which children adopt chronology, and this will be discussed in Chapter 3.

When formulating learning objectives regarding chronology, it should be acknowledged that knowledge of chronology is only a part of historical time consciousness, which in turn is only an aspect of historical consciousness. It is apparent that "knowledge of historical chronology" is sometimes equated with historical time consciousness or even with historical consciousness in general in the public debate (Stow & Haydn, 2000; Van Sas, 2009). Mastery of chronology is one of the aims of history teaching, but there are many more. The associated teaching objectives, which are rarely disputed, are objectives like:

1. Apply Western chronology and the associated concepts such as "year," "decade," "century," and "millennium."
2. Acknowledge the relative and subjective nature of Western chronology.

The assumption stated in Section 4 that cyclical, social and mythical awareness of time are also present in Western society leads to a research question relevant to history education associated with learning historical chronology, assuming that a historical consciousness of time perhaps should be viewed as an alien "exceptional category" (for students) in thinking about time. Another question is associated with the assumption that events experienced are more easily remembered than moments in time designated by calendars and years.

> Question 1: To what extent is an awareness of daily, cyclical time, social time and a "mythical" time present in students and to what extent is an orientation towards a long-term, linear temporal awareness a deviation from this in their experience?
>
> Question 2: Are references to time that make use of calendars and years more difficult to remember than references that make use of an inherent qualification of a moment in time?

Periodization

As a rule, historical time is presented in history lessons as a long-term, linear passage of time, divided into periods with individual characteristics. However, the distinction between periods is a less objectivating activity than the application of chronology. The question is whether this is always understood by students. According to the Dutch requirements for the final examination of secondary schools (comparable to the British GCSE), however, students must be able to explain that "chronological divisions are interpretative in nature."

A problem associated with learning about periodization and learning to use it is the creation of a surveyable reference framework that can be used to position historical events in the correct historical context. Research suggests that students have considerable difficulty creating a reference framework like this, let alone using it (Shemilt, 2000; Lee & Howson, 2009). The cognitive operation needed to "correctly place" a historical event or phenomenon is usually referred to as "contextualization." Research has demonstrated that experienced historians tend to do this more frequently and more easily than students (Wineburg, 1991, 1998), but it is difficult to determine what type of "period knowledge" is required to place events in the right way in context. Van Boxtel and Van Drie (2004) noted that knowledge of—what they call—certain historical "landmarks" and knowledge of "periodization" are used by students to contextualize events. Another key to a framework knowledge might perhaps be the use of "colligatory concepts"—a term introduced by the philosopher of history W. H. Walsh (1967), which he intended to refer to summative, interpretative historical concepts such as "Renaissance" or "Enlightenment."

More research is needed to clarify the role of colligatory concepts in history education.

The previous discussion focused on historians' preferences for either "neutral" or "biased" interpretative names for time periods. An educational consideration could be to choose the non-neutral, "biased" names, considering that there is no argument against this in terms of the history theory and based on the assumption that the distinction between "different time periods" is relatively difficult to learn and could be facilitated through the use of relatively non-neutral images. This can be associated with the fact that experienced events (a marriage, house move, birth of a child) are more easily remembered than a moment determined via calendar time. Research is needed to demonstrate whether a relatively pregnant non-neutral typifying of time periods is indeed helpful in the development of a historical consciousness of time. An example of this kind of research is presented in Chapter 4 of this book.

Another clue for research is the tendency defined by Zerubavel towards "interperiodic splitting" and "intraperiodic lumping." The question is whether students indeed show this tendency, and if so, which educational approach could be used to deal with this.

Teaching objectives in respect of periodization can be described as:

1. Acknowledge that periods are historical interpretations that do not just follow from "actual experienced events," while yet there might be a potential association with narrative elements present in these "experienced events."
2. Dispose of a general framework of historical periods, and master within this the specific characteristics of given periods so that the periodic framework can be used as a reference framework to allow events, phenomena, developments and persons to be contextualized in the correct way.
3. Acknowledge how continuity and discontinuity cohere within a historical development and determine that discontinuity can only be established from a retrospective interpretation based on a comparison of two moments in time, and that it is therefore not possible to say that identifying discontinuity enforces a certain type of periodization.
4. Acknowledge that different paces can be discerned in change processes, which may end up in a description of historical developments focusing on the evenemential, the conjunctural or the structural level.

Research questions:

Question 3: To what extent do students tend to perceive time periods as something actually existing, as given entities corresponding to a historical reality?

Question 4: Are imaginative-associative names of time periods (like "time of the knights," "age of steam") more easily remembered than neutral references (like 1100–1500, 19th century)? If so, are imaginative-associative temporal references supportive in creating a historical reference framework?

Question 5: To what extent do students tend to underestimate temporal differences *within* a time period and overestimate temporal differences *between* periods? To what extent do students tend to overestimate differences between time periods and underestimate correspondences between periods?

Question 6: To what extent do students tend to and are able to see that noting discontinuities rests on interpretations and to what extent are they able to distinguish between processes of change at different levels?

Generations

In the section describing a historical consciousness of time it has been argued that a "them-relationship" to people in the past is a feature of such consciousness, even when those people are still alive today and even where this concerns considering the own self as a person in an earlier, essentially different time. This rests on the distinction between the social and the historical past. The social past concerns those with whom we have an "us-relationship," even where this concerns people who (partly) lived in an earlier time, such as relatives.

There has been no specific research undertaken into the extent to which students and teachers use the "us-" and "them-perspectives" within history education, whether they are aware of this, or into the reasons why and circumstances under which they would choose one or the other perspective and the extent to which they think it is desirable to use the "them-perspective" in the context of history. Impressions from research that can be used in passing suggest that people are not normally aware of the way in which an "us-" and "them-perspective" is used (see Chapter 3). It is not usual in current history teaching to promote the use of the "them-perspective" as an explicit teaching objective. However, a teaching objective in the context of a historical consciousness of time could be formulated as follows:

- Refer to people in the past as "them," even if, in the first instance, they seem to belong to a group to with which the observer identifies him/herself in the present.

There are two associated research questions:

Question 7: To what extent do students distinguish between a historical and a social past?

Question 8: To what extent and under what circumstances do students tend to develop an "us" or "them-relationship" with people from the past?

Relics

Everything around us in this world could, in principle, be conceived as a relic from the past. The present holds numerous pasts, some recent and others more distant, in other words: there are "layers" in the modern world, defined by Pinder as "the contemporaneity of non-contemporaneous." People living with an awareness of daily, cyclical time, or with an awareness of social time of relatively short duration, or the "quotidianity" as defined by Heidegger, or those only employing Oakeshott's "practical present-past" probably have difficulties developing a sense of these deeper layers. It is perhaps comparable to the developing of a "them-relationship" with people from the past, because familiar and well-known matters in people's environments (concordant with people with whom one experiences an "us-relationship") need to be placed at a distance and need to be viewed as originating from a totally different cultural environment than the present (comparable with people with whom one has a "them-relationship"). For instance, the old church tower on the market square needs to be seen no longer as "our tower," but as the tower that strange and distant people built hundreds of years ago, when the world looked quite different from what it looks like nowadays (without cars parked on the market place, without the square being paved, and without the familiar local shops that presently shape the surroundings of the tower). Little or nothing is known from research about the associated problems this creates in teaching and learning history. In the Dutch final exam programs, the item "acknowledge how elements originating in different time periods may be manifest concurrently in the same time period by distinguishing between continuities of long and short-term duration and the contemporaneity of the non-contemporaneous" has been reserved for only the advanced, pre-university level. Apparently it was viewed by the authors of the program as a relatively difficult cognitive operation. In the German standards the contemporaneity of the non-contemporaeous appears in the requirements for grades 9 through 12, but not for the lower grades.

Teaching objectives associated with the concept of "relics" can be described as follows:

1. Explain the cultural environment as a collection of relics from various time periods from the past.
2. Acknowledge that some relics originate from recent and others from earlier time periods, and that the present therefore contains traces of different times (non-contemporaneity of the contemporaneous).
3. Apply the non-contemporaneity of the contemporaneous also to the cultural environment of time periods in the past.

The associated research questions are:

Question 9: To what extent do students tend and are they able to perceive the present world as comprising relics originating from different time periods in the past?

Question 10: To what extent do students tend and are they able to perceive the "non-contemporaneity of the contemporaneous" in a cultural environment in the past?

Anachronism

The requirement to avoid anachronisms and yet the need to use anachronisms have a particular feature within history education. On the one hand the history teacher, perhaps more so than other historians, has to explain other times in the language of his or her own time. On the other hand he/she will not be teaching the children very much if this allows them to conclude that other time periods do not differ substantially from the present. What do they learn if they are taught that "Romans were actually just normal people who just didn't have cars" (as explained by a guide to a group of children visiting an archeological theme park)? There is a strong tendency to arrogate the past in an a-historical manner in order to underline the justification for certain positions in the present, as is apparent from numerous examples in, for example, Lowenthal (1985) and Ribbens (2002). The role of historical science is to act correctively vis-à-vis certain misunderstandings and myths. To be able to place oneself in another time is a necessity, insofar as it is a precondition to being able to understand and evaluate a time period by itself, even though this is associated with the awareness that this historical ideal can never be achieved completely, as any image of the past is always perceived from a later perspective (Davis, Yeager, & Foster, 2001).

The sense of a breach between the past and present is perhaps the most fundamental element in the Western consciousness of historical time. However, it is questionable whether this historical experience of a breach and distance with the past is generally present among students. There are two tendencies that could conflict with this. In the first instance there is the tendency to extend the sense of "us" to the past ("We were then at war with Spain"), which would indicate a decrease in temporal distance. Secondly there is the tendency to believe that the thoughts and actions of predecessors were "strange" or "ridiculous," indicating the application of modern standards to the past: "typically students, particularly at a younger age, view our predecessors as morally, intellectually, and scientifically inferior and dismiss their actions as stupid or idiotic" (Foster, 2001, p. 171). People from the past are usually judged, in the first instance, on the basis of modern standards. However, this implies that, when people conclude in this way that something from the past was "very strange," the sense of distance and the awareness of an evident "unusualness" of the past do indeed come to the fore.

Modern standards also play a role in the use of the terms "not yet" or "already" ("there was no electricity at that time yet, so one had to use candles," or "the Romans already had central heating"). Using "not yet" and "already" in these contexts means that it is a normal standard to have electricity and central heating, and that it is to be considered a defect or an asset of former times if they deviate from this standard or conform to it. Research has shown that students tend to use these types of expressions (Barton, 2008). It is likely that history teachers also frequently employ them, either intentionally or unintentionally.

Other ways in which the relation between the past and present often leads to anachronisms are to be found in the search for confirming the veracity of modern visions and claims, the search for confirmation of current identity, or when searching the past for guidelines for the present (Lowenthal, 1985). As previously discussed, according to Oakeshott (1983), the study of the past with no other goal than understanding the past by itself is a particularly difficult activity that is highly susceptible to relapse "into some other engagement" (p. 28). It is as if he saw a class of students in a history lesson when writing these words. The other type of "engagement" should probably not (just) be perceived within this as the tendency of students to be interested in things other than their learning material, but also as their tendency *not* to view matters from other times as details that should only be used to acquire knowledge from those other times and to understand what was happening during those times. It is apparent from research that this type of approach is not one that students tend towards in the first instance (Wineburg, 2001). Barton (2009) has therefore argued against Oakeshott's vision and has stated that history should be used for present day aims, as it would otherwise be difficult to explain to students why history is useful at all. Thus, the question remains how "presentist" or "anachronistic" history education should be conceived.

Teaching objectives associated with the above could be:

1. Learn to not ascribe events to historical time periods if they do not belong there.
2. Avoid judging situations from the past by today's standards—avoid using terms such as "already" or "not yet."
3. Learn to not search the past for arguments to justify modern perspectives, or for a meta-historical identity of the groups people belong to or for rules of modern behavior.
4. Attempt to understand and explain the past from past circumstances and standards and to avoid passing moral judgments.

Quite a few things are known about these objectives from research (refer to Chapter 3), and they also partly occur in standard programs for educational objectives. For instance the Dutch final examination program for secondary schools refers to: "taking into consideration the time-boundness and location-boundness of interpretations and judgments by people from the past and the present" and "taking into consideration the role of values in the past and present." The following research questions remain for the development of expertise in teaching and learning history:

> Question 11: To what extent do teachers and students in history education tend to make use of or avoid using phrases suggesting that things were "not yet" or "already" the case in the past?
>
> Question 12: To what extent do teachers and students tend to use the past to justify certain claims in the present?

Contingency

The unpredictability and capriciousness of historical developments and the unintended consequences of human actions are perhaps the most important insights to which history can contribute. However, the image of the past as capricious and with little adherence to clear historical lines of development is perhaps difficult to learn, as the image is relatively illogical and incoherent. This conflicts with a natural tendency to see logic and coherence. Similar to the question for the previous concept of how anachronistic or presentist history teaching should be conceived, the question for this concept is how much logic, linearity of developments, and coherence we should admit in history in order to create an acceptable and teachable whole without losing important insights associated with the unpredictability of developments and unintended consequences of human actions.

One of the essential aspects of a temporal distance is that people are more aware at a later stage of the course that events have taken than at an earlier stage. Knowledge of the outcome interferes with the insight that a given development in the past might have taken a completely different course. A problem for learning associated with this is that in order to get a solid grasp of a certain period it may be necessary to disregard knowledge about outcomes. For instance, a student should be able to explain the behavior of the Dutch in the sixteenth century with due consideration of the possibility that the Spanish king could have won and that an independent Holland would never have arisen. Psychological research has established that people experience difficulties imagining that others do not know something that they do know; there is an unconscious tendency within people to assume that others share the same knowledge as oneself. This phenomenon is referred to as "epistemic egocentrism" (Royzman, Cassidy, & Baron, 2003). It is possible that something similar is happening when people are confronted with the past, assuming that somehow people in the past must have disposed of the outcome knowledge that later generations have. It is difficult to assess the past while consciously ignoring knowledge about later developments that people of the time did not have.

Teaching objectives associated with coincidence and unpredictability could be:

1. Acknowledge that some developments in the past can be viewed as unintended results of human actions, or the results of a coincidence of circumstances or of interference resulting from the actions of people with different intentions.
2. Attempt to understand and explain situations from the past by consciously ignoring knowledge about later developments.
3. Learn to not sketch developments of the past towards the present as a direct line characterized by progress, but as a capricious and relatively unpredictable process.

4. View one's own situation, insights, (religious) beliefs and convictions as results of a contingent, historical development that might have turned out differently; consequently view one's own situation, insights, (religious) beliefs and convictions as relative and changeable.

The associated research questions for this are:

Question 13: To what extent do students tend to and are able to explain developments in the past as unintended consequences of human actions or a coincidence of circumstances?

Question 14: To what extent do features of "epistemic egocentrism" apply when confronted with people from the past?

Question 15: To what extent do students tend to view the path from the past to the present as a direct and logical line of development?

Question 16: To what extent do students tend to and are able to relativize their present situation, insights, (religious) beliefs and convictions as historical phenomena?

Retrospective and Preview

In this chapter, historical consciousness of time has been elaborated into a workable concept for history education. The chapter started with a philosophical exploration of the concept of time. It became apparent that time is a less obvious, less objective concept than might have been initially assumed. Thinking about time leads to problems that cannot be easily solved. Perhaps this is one of the reasons why thinking about time is not particularly easy to learn. On the one hand the "objective" passage of time imposes barriers around what we can and should conceive of as a historical image: the limits of chronology. On the other hand the passage of time is perhaps something that does not exist outside of human thought; in that case time would be entirely dependent on human cognition. If that is true, it would be difficult to see why human thought about time should be bound to any rules at all. The fact that these rules nevertheless exist is a given feature of our culture, which assumes that there is an autonomous process that plays a role in the passage of time outside human cognition. Insofar as teaching involves cultural transfer, it is imperative that cultural rules about time are being conveyed to later generations.

It has been made apparent that "rules" about the consciousness of historical time can be discovered within Western historical consciousness as it has developed over the last three centuries, and in the way in which historians try to capture time in terms of narratives or emplotments. Western historical consciousness places an emphasis on the distance between the past and the present, including the in-between distances of historical periods of time. Time periods should not be anachronistically confused with each other, yet anachronistic images are a necessity to be able to say something at all about different times. There is no logical, continuous line from the past to the present, yet a certain coherence and logic is required to produce a comprehensible image of the past. Historical cognitions are often far removed from the ways in which people usually handle time. Perhaps this is why historical thinking is such an "unnatural act" as remarked by the American expert on teaching and learning history, Sam Wineburg (2001), or so "counter-intuitive" as suggested by his British colleague Peter Lee (2005).

Irrespective of how difficult the concept is, people from all cultures have an image of and thoughts about time. However, the assumption that these materialize self-evidently into the image of a principally endless timeline of successive years is incorrect. It is a "chronological illusion" (see Chapter 3) to assume that people spontaneously think in this manner. Thinking about time the way it develops more or less spontaneously displays cyclical and linear patterns. The former are related to events continuously re-occurring in people's lives, the daily cycle, the cycle of seasons and years. This is the awareness of daily time. The second are related to the fact that people from different generations live together. Parents have a different position in society than children. Parents have different knowledge and different memories. This creates an awareness of social time that ensues from the memories of the oldest living humans to their parents and grandparents, thus covering, in total, a period of approximately five generations or around 150 years. Memories of times that are more distant disappear rapidly. Beyond the last 150 years, images appear of a "long, long time ago," yet precisely how long ago is difficult to imagine without a study consciously focused on this. Without such a study, the distant past acquires mythical proportions.

It appeared to be possible to define a historical consciousness of time by means of six key concepts—in addition to and in contrast to the daily, social and mythical awareness of time: chronology, periodization, generations, relics, anachronism and contingency. A comparison with other ways of dealing with time, as they occur in non-Western cultures, demonstrated that a historical consciousness of time can be placed alongside and in contrast to an awareness of daily time with cyclical characteristics and an awareness of social time that is linear, but relatively short-lived as it is based on recollections of people from different generations. The awareness of social time is associated with an "us-perspective," while for a historical consciousness of time a "them-perspective" is used in respect to people in the past. The counterparts to the social awareness of time are the historical generation concept and the historical concept of relics from a different world. In addition to the awareness of daily and social time in our society, there may also be a somewhat mythical awareness about an undifferentiated "back then." This can be contrasted with the historical distinction of a long, linear passage of time with numerous different time periods, each with its own distinctive features, and therefore linked to concepts such as anachronism and contingency.

Objectives for history education could be derived from the six key concepts. Partly, they are obvious aims that are already the focus of most current history education. However, they are also partly aims for history education that have received little or no attention to date. They include aspects of historical reasoning that have not been extensively explored so far in studies into history education.

The research literature has wrestled with the question of what historical reasoning actually is (e.g., Perfetti et al., 1995; Van Drie & van Boxtel, 2008). The practice of history involves the application of ways of thinking and reasoning that are also used in other subjects and that therefore reflect general cognitive and reasoning skills: asking questions and formulating hypotheses, basing facts on reliable and critically evaluated evidence, providing arguments for a particular interpretation, involving counter-arguments alongside arguments, and using concepts correctly—all of these issues, which are regularly included in studies into history education as elements of "historical reasoning," are in fact nothing more than general reasoning and arguing skills. If taking due consideration of the passage of time is characteristic of history, then it is obvious that historical reasoning is a mode of thought that is based on thinking about historical time. Posing *historical* questions, basing facts on *historical* evidence, providing *historical* interpretations or arguments, must be associated one way or the other with thinking about time. Categories of this way of thinking have been detailed in this chapter.

Even without thinking about the passage of time being chosen explicitly as a starting point in studies on history education, there has been a considerable amount of research into the way in which students learn to deal with time and to reason in terms of historical time, sometimes without the studies being specifically focused at temporal experiences. These research results will be discussed in the next chapter. This will clarify which parts of the research agenda developed in this chapter have already been covered either fully or in part by existing research and which are still open for future research.

3

Developing and Learning a Consciousness of Time

Review of Current Research

No research has been undertaken specifically aimed at learning a historical consciousness of time as described in the previous chapter. However, there has been a great deal of research that may be useful to this topic. Firstly, there have been studies in developmental psychology about the ways in which time awareness develops in young and growing children. While this does not involve a historical consciousness of time, nor history education, it may nevertheless be useful to involve this research in this study, as a historical consciousness of time—as elaborated in the previous chapter—even though it is a separate category of temporal awareness, is associated with objective and subjective experiences of time in general. The psychological research is discussed in Section 1.

There has also been educational research focused on learning temporal awareness. This has occasionally run parallel to research in developmental psychology or has used the latter as a stepping stone. This research has

Images of Time, pages 95–141

virtually exclusively involved children of primary school age. There is little research about learning temporal awareness by adolescents and adults, apparently due to the fact that the prevailing idea is that temporal awareness reaches its adult stage at the end of primary school and that there is therefore little more to learn about the topic after this stage has been reached. For instance, Harner (1982) has noted that the use of temporal expressions by adolescents and adults has not been explored, as it is assumed that they have mastered the entire range. The research with primary school children is discussed in Section 2 for the period up to and during the 1960s, and Section 3 covers the period after then. This distinction has been made, as from the 1970s onwards research developed new approaches that resulted into fundamentally different insights about the ways in which an awareness of time can be learned.

From research on history teaching undertaken with older students and adults with aims other than studying the development of temporal awareness, now and then some tentative conclusions can be drawn about various aspects of historical consciousness of time as described in the previous chapter. The diverse data collected in this manner are covered in Section 4. These data have been arranged around the six key concepts of historical consciousness of time described in the previous chapter, creating an image of what is already known and what requires further research concerning the role of these concepts in history education. The final Section 5 describes the conclusions from this chapter and formulates the consequences of these for further research.

Results from Developmental Psychology Research

Research in developmental psychology has for a long time been dominated by the work of the Swiss biologist and educationalist Jean Piaget. His influence is still significant in the psychology of the development of temporal awareness. In 1946 Piaget published a book about this topic answering questions presented to him by Albert Einstein at a conference in 1928 (Piaget, 1969). Einstein wondered whether the sense of time was acquired or innate and whether it was a more fundamental than the sense of speed. This question was of course related to his theory of relativity, in which speed is more fundamental than time. Piaget dedicated a study to this with young children to see how insight into both concepts develops.

This occasion of the origin of Piaget's study clarifies two things. Firstly the supposed relevance of research among young children to adult learning: what develops in children "by itself" can be assumed to belong to the more "fundamental" and natural ways of thinking and presumably lends itself more easily to further development. Secondly, it is clear that Piaget's research involved objective time or physical time, time as a Newtonian category (see Chapter 2, Section 1), and not subjective perceptions of time. Most definitely it was not concerned with a historical consciousness of time.

Piaget noted from experiments with children that distance and space play a greater role in children's observations than time. For instance, he ran two toy trains that both left at the same time, but one was faster than the other, so the faster train covered the greatest distance, even if it was stopped first. Young children always believed that the train that had covered the greatest distance had also traveled for the longest amount of time. Age was associated by young children with someone's physical stature, so they considered it possible that one could catch up with someone else's age if only one grew enough in height. A child in this stage experiences time as becoming longer or shorter depending on the nature of the actions by which the length of time is being measured and has no awareness of a homogeneous passage of time (Piaget, 1969).

Does this mean that distance and space are fundamental and that time is actually nothing more than a construction of the human mind, as argued by "a famous school of philosophers"? Piaget (1969) argues forcefully against this position: "In this part of the book we shall try to show that nothing could be more erroneous than this Bergsonian interpretation of the psychological genesis of temporal relations" (p. 197). According to Piaget (1969) a homogeneous passage of time is as necessary for organizing an individual's life as it is for organizing the universe; it is only relatively difficult to develop a correct grasp of the passage of time. Subjective temporal awareness does receive a place in Piaget's theory when he acknowledges that different, occasionally overlapping series of events may be stored in our memories (the houses in which we have lived, developments in national politics, our holidays), which each separately display a consistent coherence, but of which it is sometimes difficult to determine the in-between relationships in time. This is why we need tools such as dates and calendars. However it boils down to the same thing—namely a problem of correctly coordinating movements, speeds and sequences.

A fully fledged awareness of time comes into being according to Piaget around 11 years of age. Similar to his general theory about the psychological development of children, his theory about the growth of temporal awareness is based on a number of more or less fixed age-bound stages. Reaching the next stage is a question of growth, not of education, and it cannot be sped up through focused instruction. Children are either able to perform certain cognitive operations, or they are unable to perform them; in the latter case teaching has to wait until the child's development has advanced far enough (Feldman, 2004). The development of a fully fledged temporal awareness coincides approximately with the transition from Piaget's "concrete operational" to his "formal operational" phase. The fact that history teaching has to wait until this crossover has occurred underlines the pessimism of some educators in the first half of the twentieth century about the feasibility of history teaching in primary schools. After all, the child leaves the primary school around the moment at which temporal awareness has fully developed.

Why is it that temporal awareness develops relatively late in children? According to Piaget the cause of this is that when thinking about time one needs to dissociate oneself from the observable present and transcend space. One literally needs to think "backwards" and "forwards." For thinking in time, it is not sufficient to simply follow the irreversible line of successive events as they consecutively take place in reality, as this would leave one in a timeless "now." One needs to move away from the point on the line where one is currently located. This mental backwards and forwards travel is a difficult, abstract cognitive operation, which in Piagetian terms, falls under the category of "formal thought" (1969, p. 259).

William Friedman is a psychologist who has intensively studied the development of temporal awareness in children and adults. Where Friedman refers to historical time awareness, this concerns memories about events from people's own lives or events that have occurred during one's lifetime (autobiographical memory). The term is not used by him to refer to the creation of an image of a historical past that has a longer duration. Friedman's historical time is therefore not the historical time described in the previous chapter. Memories are not the same as historical knowledge (Lowenthal, 1985). So it remains to be seen to what extent Friedman's research results can be applied in the case of the development of a historical consciousness of time. The section below discusses Friedman's publications, including those where he summarizes the status of research and findings from a large number of his colleagues (Friedman, 1978, 1993, 2005, 2008). It is therefore assumed here that Friedman's findings are representative of the state of knowledge in this area of psychological research.

Friedman (2008) points out that temporal awareness is an issue with numerous aspects. In dealing with time people have to come to grips with a significant number of various aspects of a problem, such as:

- Assessing durations on the basis of the logical information available (such as dates and time points);
- Estimating the duration in the absence of this information;

- Distinguishing between concurrent and consecutive events, for instance through the use of words such as "before," "after," and "during";
- Establishing a sequence in consecutive events;
- Distinguishing between the past, present and future;
- Developing insights into conventions of periodization that have been developed within our culture (weeks, months, centuries, etc.) (Friedman 1982a).

Time is a physiological, perceptual, cognitive and social, as well as a natural phenomenon, according to Friedman (1978), so one single approach to the problem will never be sufficient. Nevertheless, research into temporal awareness has often been undertaken separately in different worlds: the world of objective time (particularly arithmetic), that of social time (particularly linguistic), and that of experiential time (particularly perception and intuition). Time does not just involve adapting to a natural environment, but it requires adaptation to a social and cultural environment. Conventional time systems play a role in the latter: the clock, calendar, and history. For each of these three one has to deal with duration, the number of elements, random cycles or series or cycles based on natural laws, sequences, and repetition or return. Furthermore, there is a relationship with numbering systems, while attention has to be paid to the passage of conventional time in relation to absolute Newtonian time (Friedman, 1982b).

In terms of psychological research, a distinction has to be drawn between the perception of time and remembering a passage of time (Friedman, 1992b). This also implies an important comment to the discussion of philosophical theories about subjective time in Chapter 2, which for the greater part were concerned with the perception of time. A historical consciousness of time concerns an image of time that is in greater agreement with memory than with perception, even if memory is different from historical consciousness of time.

There are different psychological theories about the way in which people remember time (as personally experienced). Friedman (1993) refers to three main categories:

- Temporal distance theories. These are based on the assumption that memories of events in a more distant past automatically become weaker; this could be used to determine how long ago something occurred.
- Location theories. These assume that a memory is divided into "locations," where each memory is paired with an associative

context in which an event has taken place. By combining various fragments of memory we are able to reconstruct a past out of separate elements ("this has to have occurred earlier than that, because . . ."). If this is not rehearsed regularly then locations become distorted, and so we still know that something has occurred, but we no longer know when it occurred in relation to something else.

- Theories about sequences. According to these theories, a sequential code is attached to events in our memory. This type of code may also be applied to events later on in retrospect.

Friedman's research investigates how much support there is for each of these three types of theories. For instance, he conducted an experiment with groups of children who were roughly five, seven and nine years of age; these children participated in events at school that they were likely to remember (a lesson was recorded on video, so they were able to see themselves on film, and a large model of teeth was used to explain to them how to clean their teeth) (Friedman, 1991). One event had taken place one week previous to the point at which they were asked to remember them, the other event seven weeks earlier. In response to the question of which event had taken place longer ago, 30% of the five-year olds gave the wrong answer, 7% of the seven-year olds and none of the nine-year olds. It is apparent from this that a large majority of these young children had a correct insight to the time sequence and that this number increased with age. This result could be used to support the temporal distance theory, as it involved nothing more than the difference between something that occurred one week ago and something that had occurred seven weeks in the past. However, something different arose when the students were asked a number of additional questions. When asked which day of the week, which month, which season or what time of day the events had taken place, it was apparent that the largest number of children were able to remember the time of day, but not the day of the week or the month. There was little or no difference between the age groups in terms of the latter finding. There was a greater recall of the season than the weekday or the month, and here the scores were higher by age group. If the temporal distance theory is correct, then the day of the week would be as easily remembered as the time of day. However, a time point in a day or a season is related to associations, to specific circumstances in which something took place, in other words, to a context. The same also applies to the season. These findings therefore supported the location theory, although the strength of the memory probably also played an additional role in this instance.

Temporal distance theories only appear to play a role in relatively recent events (Friedman, 2005). A different recall mechanism is used for the last 24 hours or the previous week than for events that occurred five or ten years ago; the latter case requires a calculation of the correct location in time. A second experiment confirmed this conclusion. Friedman asked a group of students when an earthquake, which they had experienced nine months previously, had occurred. The students could accurately recall the time of day, but errors were made in the date amounting up to a difference of two months (Friedman, 2005). Various other studies have also demonstrated that the memory of the exact date of an event is quite weak (Taylor, 1993).

Later studies by Friedman and others have demonstrated further evidence for location theories. Friedman (1992a, 2008) refers to loose "islands of time" that initially exist in the memory of young children, as kind of separate pictures that do not necessarily fit into a single coherent time system. The loose islands of time can only be ordered by consciously imposing a separate artificial calendar system on them. Even for adults who often perceive time as a "seamless, integral continuum," it appears that the underlying psychological processes more often boil down to working with "a patchwork of representations" (Friedman, 2005, p. 155). This does not specifically involve temporal information, rather it is the contexts within which information is remembered that may eventually be placed into a calendar system with the aid of various tools. In Western culture this calendar system is a linearly ordered time system based on the assumption of the existence of a homogeneous, continuous passage of time: "an abstract, uniform, measurable dimension that stretches indefinitely into the past and the future" (Friedman, 1990, p. 103). But the notion that there is a self-evident innate awareness among people of such a long term, linearly ordered time, is referred to by Friedman (1993) as a "chronological illusion" (p. 60). Friedman argues that it is actually quite logical that linear images as such are not easily stored in memory. Continuously recurrent, cyclical patterns are far more useful and important for daily functioning. A linear pattern of ordered information about once-only events in the distant past is rarely of any practical benefit. It is therefore not surprising that human memory is only moderately equipped for this kind of information (Friedman, 1993). "The sense of an absolute chronology in our lives is an illusion, a thin veneer on the more basic substance of coincidence, locations in recurrent patterns, and independent sequences of meaningfully related events" (Friedman, 1993, pp. 61–62).

With respect to the theories on sequences, the only evidence for their validity has been found in laboratory experiments where participants had to learn to memorize lists of (for instance) words. People are often able to remember which words occurred earlier and later in the list. Whether this mechanism is also applied in daily life has not been substantiated by research (Friedman, 2005).

If we try to apply these data to learning history, we have to make some distinctions: in the first place a distinction between knowledge of events in the past that have been experienced by the individuals themselves and knowledge of events in the past that have not been personally experienced. The latter instance is the usual focus of history. Of the three types of theories only the location theory and sequence theory could play a role in learning history, as temporal distance is not relevant in the case of acquired knowledge about historical events. Temporal distance could only be applied in terms of the moment at which knowledge is acquired about a given historical event and the moment at which it has to be remembered, and can therefore bear no relevance to the time in the past at which something actually occurred. The moment of acquiring knowledge about a historical event is in most circumstances not related to the moment at which this occurred. Research into the learning of history will have to demonstrate the extent to which theories about locations and contexts and those about sequences not only apply to autobiographical memory, but also to acquired historical knowledge.

The work of Fraisse (1982), which precedes Friedman's published work, has been influential in France. He distanced himself from Piaget's theory that time is only concerned with natural phenomena about which children need to learn the correct information. People adapt in a number of ways to time; the most advanced way is that in which a notion of time is developed that enables one to discuss past, present and future coherently and objectively and to behave accordingly (Fraisse, 1982). Once again it is appropriate to note that Fraisse was not concerned with a historical consciousness of time, but with autobiographical memory. He noted with respect to objective and subjective approaches to time that disposing of knowledge about a homogeneous passage of "objective" time does not prevent us from perceiving time in a way that deviates from this. It isn't any different from perceiving space: our knowledge of real spatial dimensions does not stop optical illusions from occurring (Fraisse, 1982). This comment seems to indicate that Fraisse attached greater weight to an objectivating temporal awareness than to subjectivating perceptions, and was, in that respect, aligning himself with Piaget's rejection of the opposite stance taken by the "famous philosophers." According to Fraisse (1982), in psychology there is only subjective time, which has to be brought into alignment with objective time.

Harner (1982) has pointed out, among other things, that adult temporal awareness rests on the capacity for "temporal decentering," the ability to adopt a perspective that differs from one's own perspective at a certain moment in time. This is the cognitive operation Piaget believed was difficult to achieve, requiring a high degree of abstraction. Harner also believes that this is what makes thinking about time difficult for children, although they are able to come to grips with this in phase between seven and 11 years old. From this point on, the question is no longer how children learn to think about time, but what is their conscious perception of the past and the future. The latter, it could be added, is exactly what plays a role in a historical consciousness of time; however, about this topic we have very few data from research, says Harner.

McCormack and Hoerl (2008) have extensively investigated the phenomenon of "temporal decentering" and how this develops in young children (up to the age of five). Initially children only make a distinction between the time at which something happened and the time at which we speak about it ("event time" and "speech time," for instance if someone says, "I walked," there is a difference between the time point when the walking occurred and the time at which it is being discussed). Children are also able to choose different reference points in time at a later stage ("reference time," for instance if someone says, "I had walked," there is a third reference point between the speech time and event time). The question is whether reference times can be selected separately from events, or more generally whether time—just like space—can be viewed as being "empty." This is associated with the abstract system of clock and calendar time: adults are able to imagine 11 o'clock in the morning in the absence of any event occurring at that time. However, empty space is easier to imagine than an empty time point.

The fact that thinking about space is easier than thinking about time is also evident in language people use. In all the languages of the world that have been explored on this point, people use spatial metaphors to talk about time (Nuñez & Sweetser, 2006). Two models play a role in this process: time moving past a stationary speaker (for instance in: "the future approaching us"), or a speaker moving along a line from the past to the future (for instance in: "we are approaching the end of the year"). The usual assumption is that the future lies ahead of us and the past behind us. However, we are not entirely consistent in this respect, because we say that New Year follows Christmas (which would suggest that New Year is positioned behind Christmas), while in the model with a prospective timeline it should be stated that New Year lies further ahead of us than Christmas, and so is positioned in front of Christmas. It is apparent from this that "before" and "after" cannot be interchanged here with "in front of" and "behind" without consequences (Nuñez & Sweetser, 2006). Nevertheless, people do show the tendency to imagine time in spatial terms.

Block and Zakay (2008) have confirmed Fraisse's comments about adapting to natural time and add that people share basic notions about roughly how long ago recent events occurred and approximately how long they lasted with animals and that animals also base their actions on this. This, of course, has nothing to do with a historical consciousness of time. In addition to this they have confirmed on the basis of laboratory experiments and other experimental experiences that people base their estimates of time about past events on contextual information (location theory) or temporal distance or both.

Conclusion

Psychological research into temporal awareness has perhaps a limited validity in the context of history education, because it focuses on people's memory for events they have personally experienced. For instance, from the fact arising from research that people's memory of dates and years is poor, it can perhaps not be concluded as a matter of course that students have difficulties in remembering years and dates from history. After all, the timing of events people experience is not explicitly inculcated when they take place, while dates and times do receive special attention when learning historical information. Temporal distance theories do not relate, in the context of history education, to the distance between a historical event and the present, at the most, it could be said that they have some relevance to the temporal distance between the time something is being studied and the moment at which it needs to be recalled. Distance theories are therefore not applicable to learning a historical representation of the past. The most useful facts are perhaps the general findings from psychological research about the images of time that people develop. History, after all, also involves images of time. There are a number of conclusions which may be drawn here:

- Space is more easily perceived than time. This is because space is observable in concrete terms, whereas time is not. Confusion may arise between space and time, because people use spatial metaphors when thinking and talking about time.
- Thinking in time requires "temporal decentering," which is not easily achieved, as the most natural experience of time is the one that takes the current moment as a point of departure.
- With respect to "event time," 'speech time," and "reference time," we can conclude that the third of these predominates virtually uninterrupted when one is involved in history.

- Temporal information is stored contextually and associatively—not with the predetermined intention of remembering the passage of time—and therefore often takes the form of separate "islands of time."
- Thinking in terms of a long, linear and continuous passage of time is a cultural phenomenon that does not tally easily with human memory, which is designed to recall repetitive events at short term.
- An objectivating, logical and numerically based time system has to be imposed on the intuitive experience of time by means of explicit learning experiences.

Domain-Specific and General Educational Research Up to the 1960s

As is apparent from the above, dealing with time rests partially on spontaneous intuitions and partially on applying learned information. The use of numerical time scales and other principles of chronology belong to the latter. The research that is discussed below has studied the way in which children learn this. Although this has been partly covered by psychological research, it has been included here in the educational category, as the intention here is not to study a growth process or spontaneous development, but rather the results of learning processes. These two have occasionally been confused with each other in research, where reference is made to development when learning processes are evidently involved, such as learning to use clock and calendar time. The distinction is important because learning processes may be influenced externally, while a developmental process simply evolves as set out by nature and cannot be altered. Von Borries (1987) describes a transition in research in teaching and learning history around 1970 from "maturation theories" towards "socialization theories," corresponding with the distinction between developmental and learning processes. This is one of the reasons why educational research is discussed here in two parts: the period up to around 1970 in this section, and the period from 1970 onwards in Section 3.

An early example of a study into students' learning temporal awareness at primary schools was undertaken by Oakden and Sturt in 1922. This was an influential study in the sense that it was still being referred to as an authoritative source by authors in the 1960s.[1] The study consisted of eight written tests completed by children from eight to 14 years of age at various primary schools in Great Britain, as follows:

1. A test with various questions about time, such as: What day is it today? What's the month? What's the time? What time is it now at your house? Robin Hood was alive in 1187; was your mother alive then? How about your grandmother?
2. A test where the historical figures had to be placed in the right sequence on the basis of years. Nero was alive in 50. Dante was alive in 1312. Attila in 438. What is the correct sequence?
3. A text containing chronological inaccuracies, for instance that the sun rose and a little bit later evening had fallen; children had to identify these inaccuracies.
4. A test where children had to indicate which kind of clothing people wore in various time periods, such as "the time of Charles I," what they did at various time periods during the day, and so on.
5. A test with three images of people from different time periods, with the questions which time period this represented, how long ago it was, and so on.
6. A series of names of historical people that had to be placed in the right order: Caesar, King Alfred, Robin Hood, Charles I, Tennyson, Admiral Beattie.
7. A story that had to be retold; it was assessed whether—and if so to what extent—students used temporal information in recounting the tale.
8. An assignment to write a letter about making an appointment; these were assessed to determine whether a clear time was given for the appointment in addition to the location.

With the exception of tests 3, 7 and 8, this study explored whether or not children had *learned* something, and not whether anything had developed nor whether they were able to do something. For instance, for test 2 it was apparent that some students reasoned in this manner: Nero lived 50 years ago and Dante 1321 years ago, so Dante was the first. There is absolutely nothing wrong here with the reasoning or the mathematical skills, only the conventions for counting years had not been learned as yet. This tells us little about the question of whether students would have *been able* to learn this. However, the authors interpreted this differently and ascribed the result to something in the development of students that was hindering them from scoring well: "Children find it difficult to arrange dates correctly or to arrange historical characters under their proper dates" (Oakden & Sturt, 1922, p. 334). However, also for tests 4, 5 and 6, the matter is simply whether or not certain information had been learned. A majority of adults nowadays would probably also not be able to put the persons mentioned in test 6 in the right order.

The results to tests 3, 7 and 8 are therefore more interesting. When tracking the "time errors" in the story it appeared that certain very obvious anachronisms were easily recognized, such as that the Romans did not have any guns and that Caesar did not wear a top hat. In both instances this involves knowledge that the children must have learned. The errors that were the most difficult to identify appeared to be the errors against conventions in the numerical time system: February does not have 30 days (the 30th of February was mentioned in passing as a date), and 58 BC is three years before 55 BC and not three years after. Also in this case it applies that children must have learned how to deal with these issues. But it also tallies with the observation in Section 1 that chronology has to be imposed artificially on temporal experiences and is relatively difficult to acquire.

Something similar to this was also apparent in test 7, the retelling of the story. Temporal information was recounted significantly less often than information about the story plot. The phrase "around six in the evening" scored highest among the temporal information that was reproduced and the year and month reported in the story had the lowest scores. The fact that the time of day was remembered more easily than the month and year concurs with Friedman's findings. However, it is important here that this apparently also applies to remembering a historical story, and not just to remembering one's own experiences. This would mean that some findings from psychological research not only apply to memorizing autobiographical information, about also to remembering historical information. More confirmation of psychological research was found in test 8. Based on this research, it was to be expected that the location of the appointment would be mentioned more frequently than the time when writing the appointment letter, and this was indeed the case. The preference for space over time that was shown in psychological research was therefore also manifested here. (It should be noted that the psychological research we refer to here was carried out after the publication of Oakden and Sturt's article, so correspondence and confirmation can only be established with hindsight).

Oakden and Sturt (1922) did not recognize that tests 3, 7 and 8 were different categories of tests in respect to the others. They were even inclined to attribute greater significance to the results of tests 1 and 2 with respect to the question of whether history teaching was possible: "The knowledge tested under the first two heads is clearly an indispensable preliminary to any profitable study of history" (Oakden & Sturt, 1922, p. 311). Their conclusions about the possibility of history teaching to young children agreed with the pessimism we referred to before. A fully fledged awareness of time develops between the ages of 11 and 14, and no attempt should be made to teach history any earlier, said Oakden and Sturt. They noted a remarkable kind of "leap" in development around the age of eleven. This points to something like the phases in child development distinguished by Piaget. However, Piaget's work was published later and would only become known in the English-speaking world via translations in the 1960s.

The work carried out by Kelty (1925), an educationalist and history teacher, with children at primary schools around the same time was completely different in nature. She was one of the first researchers in the United States to focus on the development of a professional study of history education based on empirical results (Barton, 2004). Her work was a critical addition to the sparse research into temporal awareness existing around 1925. Kelty did not understand why it would be impossible to teach students to understand years such as 1492, as soon as they were able to deal with numbers above one thousand. If they were not able to do this, it simply meant that they had not been taught it. She also noted in addition to this that an understanding of a system of years was of limited use, as the images formed by children about given time periods were more important; these images would initially be separate from each other, and would only be able to be ordered later within a numbering system. These comments agree with the results of the psychological research discussed in Section 1—research which was still to be started at the time Kelty was writing this.

In her own research into temporal awareness Kelty wondered whether students were able to understand temporal expressions employed in youth literature intended for them. She firstly created a list of frequently occurring temporal indications for indeterminate, determinate and comparative times based on one hundred popular children's books for children aged five to nine. She defined determinate temporal indications as indications referring to a specific time, such as a minute, quarter of an hour, hour, month, season, year, or century, as well as those that refer to a particular time, such as Christmas, New Year's day, Easter and Valentine's day, and those referred to by a specific date. She defined indeterminate references as expressions such as "there was once" and "a long, long time ago." Comparative expressions were those identifying a moment in time in relation to another, for instance "xx ago" (in relation to the present) or "before" or "after" (such as: after the Civil War, so two moments in the past in relation to each other).

She presented children aged nine to 12 with five tests to assess their knowledge of these concepts. One test included a text where children had to identify the four ways in which time had been defined. The students struggled with this, which Kelty ascribed to reading problems, as the children did seem to be able to cope well with the same time definitions in other tests. This broadly applied for all words that had been extracted from the children's literature. The reading problem highlighted by Kelty in the first test could explain the poor results of other relatively "linguistic" tasks that had been presented to children by other researchers.

Students scored a poor result on the question of what it meant to be alive in 1924, which boiled down to explaining the Christian calendar. Questionings about assigning years to the correct century (e.g., 1672 to the seventeenth century) also had a relatively poor result. Both issues belong to the arithmetic-chronological aspects of temporal awareness, which were shown by the psychological research described in Section 1 to be artificial and relatively difficult to access. When asked whether they preferred determinate or indeterminate time references (what do you like best: a story with the words "a long time ago" or a story with "1453") the majority of students answered that they preferred the indeterminate time reference. This also agrees with the data from the psychological research discussed earlier.

Kelty's research was rather an exception among the educational studies into temporal awareness conducted up to the 1960s. Other studies mostly reflected Oakden and Sturt's approach and the conclusions drawn roughly agreed with those in their work.

Harrison (1934) reported a study where fifty temporal concepts had been presented in verbal interviews to children aged from five to nine. The answers were divided into categories from "full understanding" to "total confusion" (with some in-between categories), as well as a category of "don't know" answers. The concepts used were not reported, nor were the criteria used to classify the answers or the interview procedure. The total percentage sum of answers in the category "confused," "incorrect," and "don't know" appeared to decrease with age: 56% for 5–6 years, 45% for 6–7 years, 35% for 7–8 years and 23% for 8–9 years. However, the reporting of this study is so poor that it is difficult to evaluate the significance of these findings. Harrison (1934) believed that an understanding of temporal concepts was an essential condition to developing a sense of historical succession and cited Oakden and Sturt, who believed that this sense only develops fully around the age of 13. However, she also was of the opinion that more effort needed to be made to "fill" time periods with events that are interesting to children. She based this advice on her experience with children (Harrison, 1934). Harrison's advice therefore was to develop something like "islands of time" in the style of Friedman (1992a, 2008): periods with interesting events, which could be ordered in a chronological system at a later stage.

Ten years later Kopple Friedman (1944) (not to be confused with William Friedman) reported a test carried out with approximately 700 students from primary schools, all age groups. Students in the lower age groups were assessed to what extent they understood the conventional time system (minutes, hours, morning, afternoon, days of the week, months of the years, counting years) and students in the older age groups were assessed for what they understood about temporal information, such as that occurring in history lessons (centuries, historical years, timelines). The results of these tests were described in broad terms. For instance, it is not clear whether this was a written or verbal test and how the results were evaluated. The conclusions agreed with Oakden and Sturt's research: students can manage conventional time systems satisfactorily at the end of primary school, whereas their mastery of the historical and chronological system at that age is still inadequate. Teachers were warned to take into consideration that much is learned outside school in terms of the first temporal system; they were also advised to provide more structured instruction for the second system.

The study reported by Bradley (1948) was more or less a repetition of Oakden and Sturt's with a number of similar tests, and largely similar results. The reason why Bradley wanted to repeat the study was that he believed that some of the conclusions from the previous research were unexpected, with too little supporting evidence, in particular the finding that a sudden acceleration occurs around the age of 11 in the development of an understanding of time. Bradley noted in his own research that this was not the case, but that a gradual growth seemed to occur across the school years. The general conclusion that a fully fledged temporal awareness only occurs at the end of primary school and that therefore history education should only start after this point, however, remained unaffected by Bradley's results.

In 1963 Jahoda summarized the results of existing research and applied them to history education. He referred explicitly to Piaget, whose work had penetrated as far as the English-speaking world by that time, and also referred to various aspects associated with temporal awareness, such as language and math. Children often learn words before they have a full understanding of their meaning, according to Jahoda (1963), so their actual understanding of time lags behind their mastery of the words they use to express time references. Other mechanisms are brought into play for recalling the last 24 hours or the last month than are used for recalling something that occurred five or ten years ago, according to Jahoda (1963). Chronological awareness is required for the latter. Children who do not possess this will be unable to order events that occurred longer than six months ago. Jahoda viewed this as a precondition for learning history. He summarized all the available research results in an overview and subsequently confirmed in broad terms the conclusions that had previously been drawn by Oakden and Sturt, and Bradley. He applied the research results to history education to a greater extent than had been done previously, and it is perhaps somewhat obvious that he felt pessimistic about the feasibility of history teaching to children at younger ages. According to Jahoda (1963), students do not distinguish between periods before the age of 11: "Nearly all seem to have 'lumped' all historical periods together, giving them the general attributes of savagery" (p. 97). This provides, by the way, perhaps a good example of what we have argued in the previous chapter about the presence of something like a "mythical awareness of time" (an undifferentiated mythical past) in modern society. An understanding of the differences between various periods gradually develops after this, says Jahoda, but this is a slow process. He even referred to work by researchers who believed that a full understanding of different periods only develops around the age of sixteen. Embracing a Piagetian view on development implied that this could not take place any earlier: students were simply not in the stage of development to be capable of doing this.

A study carried out by McAulay (1961) with seven and eight year old students at three primary schools in Pennsylvania brought him to diametrically opposite conclusions. He presented students with questions about the awareness of daily time (hours, minutes, and so on, but also about grandmother's and mother's age and how these compared with each other by asking students to draw a line with the length of their ages) and their historical awareness (Who came first: the whites or the Indians? Draw something about their lives. Tell something about Washington's time. Tell something about Lincoln's time). The children appeared to be able to answer the "historical" questions better than questions about their awareness of daily time. One of his conclusions was: "The child seems to be able to project himself into the past as regards persons, places and events, but probably more efficiently with events" (McAulay, 1961, p. 314). This latter addition was based on the fact that students were better at answering questions about the Indians and whites than about Washington and Lincoln—which is not entirely surprising.

McAulay's study was heading towards a separate distinction of historical temporal awareness, differing from insights into clock and calendar times, which probably had little to do with the former. However, the assumption that a mastery of clock and calendar time was a precondition for an optimal appreciation of historical time dominated research until around 1970. It is also included in a German handbook of expertise in teaching and learning history written by Ebeling, which was first published in 1965 and reprinted in 1973, as is apparent from a diagram that includes the phases in the development of temporal awareness.

Conclusion

A conclusion drawn by virtually every study during this period was that students learned to handle conventional time systems—that is clock time, calendar time and years—as well as the vocabulary used in daily life for time references, by the end of primary school. Research in this period was unable to provide a consistent image regarding other issues. This was probably due to the fact that a number of issues were confused with each other, or summarized differently by various researchers:

- Firstly there is the question of whether acquiring an awareness of time is a learning process or a developmental process. Some researchers assumed the latter, others the former. For instance, Kelty believed that learning eras should be possible for students who understood numbers up to one thousand. All the other studies into the mastery of chronology that were carried out became unsettled when seen against the backdrop of this observation. If students appeared not to be able to perform something, that meant no more than that they had not learned it—it did not mean that they could not have learned it.
- In the second instance there is the issue of linguistic and word knowledge and temporal awareness. Although it was concluded in general that students at primary schools could master conventional concepts of time, researchers like Jahoda were warning that word knowledge does not mean that the world behind those words is being understood. He referred in this respect to a chronological awareness in which mathematical skills seem to be involved. The psychological research discussed previously has shown that this is an unruly thing to human brains, which is not easily grasped. It is therefore not surprising that primary school students made a lot of errors with it. Another question was whether this prevented them from learning something about the past. Jahoda, among others, believed that this was the case: students mixed up all periods into a single heap and believed that everyone in the past was a barbarian. However, other data suggest that mastery of linguistic and mathematical chronology is not a precondition to learning history at all. But most researchers in this period took it for granted that it was a precondition and made this their point of departure.
- Finally, the question of whether of a historical consciousness of time is dependent on mastery of clock and calendar time. Some researchers even based the design of their research on the as-

> sumption that it was obvious that telling time on a clock and the ability to place the months of the year in the correct order had to be in place before historical insights could be developed. For instance, in the study by Kopple Friedman (1944), the youngest students were only asked questions about clock and calendar time, and the older students were asked about historical time. However, it was clear from Oakden and Sturt's study, using a similar design, that children were able to retell a story about the past, including time references—although years and other mathematical references for time were not media that were readily used by the children. McAulay's research demonstrated that children aged seven and eight had images of various worlds from earlier times. These results can be easily explained on the basis of a location theory of memory (see Section 1). This also applies to children's preference for spatial references over temporal references as was apparent from the research described above.

Researchers from this period did not undertake any studies into historical consciousness of time; rather their work was focused on temporal awareness in general. No distinction or barely a distinction was made among the various types of human temporal experiences. Piaget's position that there is but a single type of temporal experience that is standard, namely that of a homogeneous passage of natural time, was shared by virtually everybody in this period. This created a relationship between learning to use the clock and the months of the year and learning history, even though these are perhaps very different categories. A good insight into arithmetical chronology was more or less viewed as a precondition for a historical consciousness of time. Only McAulay's study was heading in the direction of defining historical time as a separate category. The pessimism about the capacity of children in the early years of primary school to learn history results to a significant extent from the fact that this distinction was not generally made.

One conclusion that may be drawn from the research results out of this period is that some results from psychological research concerning autobiographical memory probably also apply to remembering historical information. The same applies to the artificial character of the conventional system of clock time, calendar time and eras, which are not easily accessible to people. It also applies to the location theory of memory: information is remembered most easily in contexts, without specifically referring to temporal dimensions. The fact that researchers of this period did not draw these conclusions is logical: the psychological research on which they are based was only conducted for the main at a later stage.

Domain-Specific and General Educational Research After the 1960s

A new phase of research into the development or learning of temporal awareness was introduced in the 1970s. Up to that point the prevailing assumption had been that each learning about temporal awareness belonged to a single category or at least a single learning path, so that learning to use clock time and placing the days of the week and months of the year in the right order, needed to precede learning about history. From the 1970s onwards, the numerous diverse aspects of the concept of time were taken into consideration. For instance, Poster stated in 1973 that although a lot of "good research" had been conducted (referring to Oakden & Sturt, Piaget, Harrison and Bradley), there was still room for something new, as there were a number of different types of time: social time (the time governing people's lives), literary time (which can be shaped at will in books), personal time (memories about one's own past), physical or clock time, as well as historical time. The latter he defined as time as shaped by historians. The focus of the research, according to Poster (1973), had always been on children acquiring an understanding of physical time (by which he apparently also meant clock and calendar time, as well as conventions of chronology); yet historical time is much more than this. He saw the relationship between acquiring an understanding of physical time and historical time the same way as previous researchers—in other words, that the concept of physical time is a precondition to understanding historical time. But according to him, it is a necessary but not sufficient condition for the development of the sense of the past that is required for history.

The rest of Poster's (1973) article consists of a discussion of the research literature from which it would be apparent that there should not be a lot of optimism about the rate at which children develop a truly historical consciousness of time. He refers to a number of sources including Hallam (1970), according to whom students of history only reach Piaget's formal-operational state (the stage at which the capacity of abstract thought arises, which in Piaget's view starts to occur around the age of twelve), at around sixteen years of age, the reason for this being that the world of history is too far distant from the student's own daily world.

Thornton and Vukelich (1988) also viewed historical time as a separate category by itself. They provided an overview of what research had yielded in terms of children acquiring an understanding of clock time, calendar time and historical time. Each of these three was sketched as a separate development, posing the question about their mutual relationships. In the development of the concept of historical time, which they defined as "the ability to depict a person, place, artifact or event in the past using some form of time language" (Thornton & Vukelich, 1988, p. 70), they made distinctions between the following stages:

- 3 to 5 years: distinguishing sequences in daily events; distinguishing age difference with people who are older.
- 6 to 8 years: use of numbers in two ways—indicating how much older others are, and attempts to use numbered years (both often incorrectly).
- 9 to 11 years: distinguishing between different time periods; in this phase a student should, for instance, be able to say "in the colonial age, some 300 years ago" (in reference to the history of the United States).
- 12 to 14 years: developing an adult time system, with terms such as century, the capacity to place numbered years within centuries, and so on.

Despite the fact that Vukelich and Thornton (1990) saw historical time as a separate category, they believed that all these phases had to be completed as a precondition to learning history, that is, they reflected this prevailing opinion within research: "Once a student has mastered time term hurdles of this type, he/she is often considered adept at using temporal skills of a historical nature" (p. 23). In this way the definition of a historical consciousness of time remained linked to the difficultly accessible, numerical, conventional time system used by adults in the Western world, a system that according to psychological research does not fit in easily with the way in which people normally use their memories. Research results showing that students were quite able to recall stories of the past, as those of Oakden and Sturt's tests 3 and 7 or McAulay's results, were not as such considered.

In the 1990s Barton and Levstik (1996) distanced themselves emphatically from existing research into the development of temporal awareness in children in relation to their ability to learn history. They thought that empirical research had still been unable to demonstrate any connection between the development of a general understanding of time and the ability to learn history. Time is a cultural construction with different aspects and there is no reason to assume that these cannot be learned independently of each other: "We see no reason to think that a child must be able to name the months of the year before he or she can recognize that a picture of colonial America is older than one from the 1950s" (Barton & Levstik, 1996, p. 422). Existing research was based on the adult world of chronology, dates, knowledge of time periods and famous people, and consequently revealed what children had and had not mastered of that adult world. Instead, Barton and Levstik wanted to investigate what children *were* able to do without departing a priori from the adult world. Their research employed nine images from daily life in the United States: one from the eighteenth century, three from the nineteenth century, two from the twentieth century prior to 1950, and three from the twentieth century after 1950 (images in appendix to their article, pp. 447–451). By using images they avoided the problem of linguistic and word knowledge that had often frustrated previous research. Knowledge of time periods and historical facts did not play a role *per se* in the design of their study, as the images did not refer to well-known historical events or conditions.

They presented the images to children from all age groups at primary schools (including kindergarten) in personal interviews. Two images were firstly presented that had to be separated into "recent past" or "distant past" and subsequently the other images were presented in sequence, which then had either to be placed in between the ones already present, or before, or after. Children had to think aloud during this process about their decisions, so their mode of reasoning about the passage of time became clear. All age groups appeared to make some distinction between the past and now, including the five year olds. The youngest often separated the images into two or three groups, defined, for instance, as "real old," "just old," and "close to now." The in-between temporal distances *within* groups for the nineteenth century, pre-war and post-war twentieth century images were not often observed or were assessed incorrectly. Older age groups were able to separate a greater number of different times and numbered years were also mentioned, which might, however, be incorrect, occasionally by hundreds of years. Children frequently used associative names for times, such as "time of the cowboys" or "wartime." In the two highest grades of primary school they sometimes used historical information, which is not surprising, because students in those grades in the United States have history lessons. Even the use of conventional chronology by these groups became slowly more accurate, although it certainly was not without errors.

Research in Great Britain showed similar results (Stow & Haydn, 2000). Students from age six onwards appeared able to place images in sequence, and students nine years of age could correctly group images from five different periods and place them in the correct sequence (Roman, Tudor, 19th century, the 1940s, and the 1990s). It was apparent that associations with periods develop more quickly than the use of a mathematical chronology: "The understanding of period seemed to develop more readily than a confidence in dates" (Stow & Haydn, 2000, p. 89).

Barton and Levstik (1996) concluded that children have the capacity from a young age to distinguish between changes in time and that within this process they make use of material and visible factors: horse and cart preceded the car, clothing that looks "primitive" in comparison with current attire must be old ("this has to be a long time ago, as these people are walking around in rags"). Barton and Levstik's advice on the basis of this study was that education needs to focus more on building a sense of the past and the present, which young children already possess, rather than assuming that it is best if they do not learn history because they have not developed a significant amount of temporal awareness.

Barton has undertaken similar research in follow-up studies with American and Northern Irish children, including time periods that were longer ago than the last two hundred years (Barton, 2001, 2002). These studies provided similar results, albeit that Northern Irish children had different perceptions of change and progress in history than American children (more about this in the next section). After the follow-up studies, Barton produced an analysis of the strategies that students apply when asked to place images into particular time periods. He distinguished four strategies:

- The use of knowledge about objects, people and events (styles of cars, clothes, armor, something that looks as if it is "wartime").
- The use of knowledge from experience in one's own social environment, for instance, my father learned to drive in a car like that, and he is this old, so this is from that time (children were able to make estimates like this with a great sense of accuracy).
- Distinguishing progress and development: bigger and taller buildings, so more recent, better clothes, so more recent. In general, if it looks more like the present it must be more recent.
- Anchoring and adjustment. If an image was identified as "now" or "nearly now," the previous images were dated by constantly subtracting a fixed number from the current year, for instance, ten or a hundred years. This lead to "disastrous results," for instance one estimate for the Stone Age was 120 years ago (Barton, 2002).

These strategies belong to at least three separate categories: distinguishing sequences, grouping images, and estimating temporal distances. Images could be correctly grouped, but the sequence then could be incorrect; the sequence could be estimated correctly, but then the temporal distance was incorrect (Barton, 2002). The use of a mathematical chronological system was dominant in almost none of the cases; there was therefore no reasoning on the basis of numbered years to determine a sequence, but rather the information in the images was used in order to estimate sequences and temporal distances. The eventual estimation of a numbered year or temporal distance followed from this. The conclusions by Barton appear to agree with Friedman's findings regarding islands of time and the location theory of memory. This is once again a fact that indicates that these do not only apply to autobiographical memory, but also to the construction of an image of historical time. The same also applies to the artificial and relatively inaccessible character of mathematical chronology.

Some research appeared in Great Britain around 2000 focusing on temporal awareness among primary school children, which had been conducted by researchers who apparently were unaware of Barton and Levstik's work. Hoodless (2002) once again emphasized how much of a key role language played in the development of temporal awareness, something that had not been confirmed by Barton and Levstik's study using images. Hoodless's work concentrated on the role of stories in children's temporal awareness. Her research demonstrated that children are easily able to manipulate time within the context of a story. For instance, she discusses a story about a dream that appears to last a long time, although on waking only a short period of time appears to have elapsed. Children aged between six and seven appeared able to function adequately in terms of mastering the vocabulary referring to time, of distinguishing between sequences, of awareness of different time periods running in parallel (the dreaming child and the mother), and of being able to make leaps in time (flashbacks and flash forwards). Hoodless's (2002) conclusions appear to support the location theory of memory given the fact that stories, like images, may provide conceivable contexts for islands of time, which can easily coexist and be put together at a later stage into a coherent chronological order.

Hodkinson (2004) commented that there were a significant number of contradictory outcomes in the research into temporal awareness published up to that time. He noted that the acquisition of temporal awareness had been interpreted by some authors as being dependent on the child's mental maturity, and by others as dependent on the acquisition of language or numeracy skills. The research questions he subsequently posed were relatively fundamental: Are children in primary school capable of learning a date system? Can the concept of a "century" be used? Does the use of subjective indications of time promote or hinder the development of temporal knowledge? He collected data from four groups of eight to nine year olds and one group of nine to ten year olds at one primary school during an entire school year. During this study some groups received "normal" lessons and others "special" lessons. It is not entirely clear what the special approach contained, but in every lesson of the "special" approach systematic attention was paid to the time aspects associated with the topic. The groups receiving the special lessons appeared to be able to understand the conventional date system and chronology better than the other groups. Hodkinson concluded that learning this material is not related to mental maturity or a stage of development in which students are, or are not yet ready; it is simply a question of good teaching methods. The children's abilities to handle time are commonly underestimated a great deal.

Conclusion

Research in this period led to the link between a historical consciousness of time and general temporal awareness being detached. Time is an issue with numerous aspects, of which historical time is only one. The conclusions regarding the question of how a historical consciousness of time corresponds to other forms of temporal awareness were not concurrent. Some researchers maintained the view that though it might be true that a historical consciousness of time is much more than clock time, calendar time and chronology, the mastery of these is still a precondition to acquiring a historical consciousness of time.[2] Others were of the opinion that conventional clock and calendar time on the one hand and historical time on the other were independent; historical consciousness of time involves a sense of other times, a sense of the changes that have taken place over the course of time, and of differences between diverse periods. When considered in this last sense, it was apparent that children are able to develop some historical consciousness of time at an early age. However, the question of how numerical chronology can play a role in this remains out of the picture, while it is also clear that students do attempt to apply this kind of time calculations around the age they are confronted with them (from approximately age ten), sometimes leading to an improved accuracy of their estimates. It is understandable that Levstik and Barton set themselves against previous research based on the priority of numerical chronology and adult standards of temporal awareness, which did not clarify the things that children *were* actually capable of doing. But the question is whether they went too far in this when they said: "Historians have no monopoly on the use and interpretation of the past, nor are their thought processes somehow superior to anyone else's" (Levstik & Barton, 2008, p. 111). This, after all, would mean that within the historical discipline no standards could be set regarding what students should have to learn about historical time and that any way of thinking about the past and present is equally acceptable.

One issue that appears to have been definitively resolved in this period is the question of whether acquiring a sense of time involves a developmental process or a learning process. The development of temporal awareness is a learning process, excluding perhaps the earliest pre-school phase of development. The fact that this involves a learning experience is not surprising given that time is largely a cultural phenomenon. Virtually everything that children have to learn about it rests on agreements between people; these agreements differ from one culture to another, as has been shown in Section 4 of Chapter 2. The statement that the development of time awareness is a learning process may lead to more optimism about whether students are *capable* of learning it. The work by Poster (1973) and Thornton and Vukelich (1988) represented a transition stage in this respect. Hodkinson (2004) in particular demonstrated what could be achieved if appropriate teaching methods were applied.

It is highly likely that the use of appropriate teaching methods does not imply the use of a chronologically ordered curriculum, starting at the "earliest times" and gradually working through to the present. This does, at first glance, appear to be the most obvious option, and virtually all history teachers assume that proper history teaching should fit together like this. However, research suggests that this does not promote temporal awareness very well. Barton (2009) notes that the successive treatment of single chronologically ordered periods "flattens" the temporal experience, as only one single reference in time is being presented simultaneously, so there is only the opposition between "then" and "now," and not between different "thens," which might result in every "then" being perceived as at more or less the same distance from now. Temporal awareness may be better advanced by drawing comparisons between various time periods, with frequent leaps backwards and forwards in time. He refers to the curriculum in Northern Ireland where the Mesolithic period and the nineteen fifties are covered in a single year, after which the curriculum turns back to the Vikings. Instead of the children becoming confused, they become accustomed to using different reference points in time, and their temporal awareness is enhanced (Barton, 2009). Vivienne Little (1990) has also highlighted studies that have shown that a chronologically ordered curriculum is not a necessary nor a sufficient condition for developing a historical consciousness of time.

Finally, the research results from this period seem to confirm the conclusion—like the earlier research results—that some outcomes from psychological research regarding autobiographical memory also apply to remembering historical information. The location theory of memory developed in psychological research also appears to apply to remembering historical information. The insight by psychologists that the conventional system of clock time, calendar time and time periods is not easily accessible and has to be imposed artificially to temporal experiences concurs with diverse research results, including those by Kelty (1925) suggesting that children prefer indeterminate temporal references rather than determinate references. For instance, the use of associative temporal references noted by Barton and Levstik (1996), such as "time of the cowboys," is preferred to the difficult and often error-prone use of numerical chronology. And Hoodless (2002) found that students can move from one time to the next within the context of a story with ease without feeling the need to precisely "calculate" that time.

Diverse Data from Educational Research About the Six Key Concepts of a Historical Consciousness of Time

The discussion in the previous sections is relevant, virtually in its entirety, to the first key concept of a historical consciousness of time as described in the previous chapter: chronology. This principally concerned students of primary school age. This section will explore which research results are applicable for older students in terms of chronology, and which research results can be used for the other five key concepts of historical consciousness of time as described in Chapter 2.

One of the studies that can be utilized is a large-scale study carried out in 1995 and 1996 among youngsters aged from 14 to 15 across thirty European countries. Around 32,000 students received an extensive form with questions about their experiences of history, their historical awareness, and historical knowledge. The results were published in 1997 in two volumes, along with a CD-ROM entitled *Youth and History* (Angvik & Von Borries, 1997). Data have been drawn from this study that were relevant to the six key concepts described in the previous chapter.

Chronology

How well were the 14 and 15 year olds in the *Youth and History* survey able to cope with chronology? Did the results confirm that mastery of chronology is completed around the end of primary school age? Four tasks included in the questionnaire for *Youth and History* requiring students to place events and phenomena in the correct time sequence could help shed a light on these questions:

- Put into the correct time sequence: a. work in factories, b. trade between cities, c. agricultural work, d. hunter gatherers, e. automatic mass production.

- Put into the correct time sequence: a. the Russian Revolution, b. the Second World War, c. de-colonization of Africa, d. creation of the United Nations, e. large global crisis (the Wall Street Crash).
- Put into the correct time sequence: Images of ships. Students were presented with five drawings, including a Medieval cargo ship, a Greek trireme, a steam paddle boat, a steam mail boot, and a Viking ship.
- Put into the correct time sequence: Images of pairs of individuals dressed in different clothing. Students were presented with five drawings, including an eighteenth century couple with a hooped skirt and wig, a nineteenth century couple with a top hat and trailing dress, a knight and a young maiden, a couple dressed in Renaissance style clothing, and a Roman couple (man dressed in a toga) (Angvik & Von Borries, 1997).

For the first question, an average of 74% placed the phenomena in the correct time order. The majority of errors were made with the factory work and trade between cities, where the order was reversed. The results of the second question were much worse: 27% correct. The de-colonization of Africa was placed first by a large number of students, and the global economic crisis was placed last by a significant number as well. In general the results of this task demonstrated a confused picture. For the question about ships the average score was 55% correct. A very large majority (almost 80%) of students placed the Viking ship first, presumably because this looked like the simplest vessel in the images: a single sail and a relatively small vessel. The trireme had two sails, and the cargo ship multiple sails. The steam ships could be recognized from the funnels (although the artist had not included any smoke plumes, so the students had to look closely at the image—the steam paddle boat also had sails, but these were folded away). The clothing images resulted in 73% correct answers, with the largest number of errors being made for the eighteenth century and Renaissance couples. The Romans were most easily recognizable: 85% placed them first (Angvik & Von Borries, 1997).

These results appear to confirm a number of conclusions stated above. The three questions containing images were answered considerably better—either due to the images or the "imaginable" lifestyles—than the second question about historical events, which could only be answered correctly on the basis of specific knowledge about the chronology, preferably through the use of years. This agrees with Barton's (2002) conclusions about the importance of material, visual factors, with psychological insights about associative contexts, and the relatively difficult accessibility of mathematical chronology. Furthermore, the answers to the question about ships agree with Barton's conclusions about "progress" as an important key factor: more developed and complicated was obviously seen as more modern, and therefore the simpler Viking ship was noted as being older than the Greek trireme. In general this may provide support to the view that results from research among primary school students also have some relevance to the situation among older students, although further research is required to know how relevant they are.

Periodization

The concept of periodization is associated with the fact that people allocate certain characteristics to certain periods of time or have particular associations with various periods. A number of questions from the *Youth and History* survey are relevant to this, in particular questions about the associations students had with "The Middle Ages," "The Period of Colonialism," and "The Industrial Revolution." The relevant data from the survey have been summarized in Table 3.1 (Angvik & Von Borries, 1997).

TABLE 3.1 Results from *Youth and History* on Associations with Periods

Averages of Likert-values 1–5 (totally disagree, disagree, undecided, agree, totally agree)	
What do you associate with the Middle Ages?	
A dark and superstitious era.	2.88
A glorious period for my country.	2.70
The time when the great cathedrals were built.	3.13
The period when Nobility, Church and King dominated peasants.	3.63
A romantic, adventurous period of knights and ladies.	3.04
A time of struggles between Church and King in many European countries.	3.40
What do you associate with The Period of Colonialism (Up to the First World war)?	
A period of great adventurers (Columbus, Vasco da Gama, etc.).	3.45
Christian mission outside Europe.	3.00
Worldwide empires of some European nations.	3.29
Starting period of European exploitation of overseas peoples and countries.	3.49
European work for progress in other countries.	3.01
Contempt for and prejudices against other cultures and colored people.	3.48
What do you associate with The Industrial Revolution?	
The beginning of environmental pollution.	3.48
The source of improving living standards.	3.56
The invention of better machinery.	4.01
The accumulation of big capital stocks.	3.46
Overcrowded and ugly towns.	3.26
Struggle between workers and factory owners.	3.58

Source: This table was created by the author based on data from the database on the CD-ROM Youth and History.

It is noticeable that the associations with the period of the Industrial Revolution are more pronounced that those with the Middle Ages and the Colonial Age (for the Industrial Revolution there was a average of 22% "neutral" choices, compared to 27% and 29% for the Middle Ages and the Colonial Age respectively). The most pronounced association of all was that of the discovery of better machines during the Industrial Revolution: an example of concrete, visible progress and development (in contrast to an equally concrete and visible decline—"ugly cities"—which scored relatively low). Furthermore, stereotypical situations such as repression and war were, relatively speaking, quite pronounced: repression of serfs in the Middle Ages, exploitation of non-Europeans during the Colonial Age, the battle between Church and State during the Middle Ages and the struggles between workers and factory owners during the Industrial Revolution. Important, but less clearly visible, background developments (such as the development of non-European regions, the Christian mission, accumulation of capital, start of environmental pollution) were not noticed as much proportionally.

These data confirm Barton's (2002) impressions about the importance of visible and material elements and the idea of progress in history. The stereotypical "repression" and "war" could be viewed as a confirmation of Jahoda's (1963) belief that the whole of the past is viewed as one set of misery and barbarism. These impressions are globally speaking in some agreement with each other. However, not much more can be said about the validity of these theories on the basis of these data.

Periodization is related to change and continuity, which are inseparable concepts in history. They are often quoted together in definitions of key concepts that are of importance in teaching and learning history (Seixas, 1996; Lévesque, 2008). However, educational research has concentrated on how students deal with the concept of *change* by itself, and not on the way in which students deal with change and continuity in association. Some definitions of important key concepts for teaching and learning history also only mention the concept of change and do not consider the interplay between change and continuity (Lee, 2005; Limón, 2002). Furthermore, what little research there has been has also focused on the *causes* of change, soon ending up with causality, a different concept that also plays an important role in history education.[3]

Research into the definitions used by students for the concept of change demonstrates that changes are often seen as being equivalent to events, and periods in which little or nothing changed are often seen as times when, according to students, "nothing happened" (Shemilt, 2000). According to Shemilt adolescents have a fairly "binary image" of the past as an "events space": something happened/changed or nothing happened. If this is true then it would mean that students have very little or no awareness of the complex association between change and continuity or of the foundations of change processes, such as structural, conjunctural and evenemential—as distinguished by Braudel (1966).

The way in which changes are perceived by students becomes slightly clearer from an item from the *Youth and History* study where students were asked how important certain factors were in bringing about changes. The results have been summarized in Table 3.2 (Angvik & Von Borries, 1997).

Given that many students will be tempted to answer a question like this positively (the factors referred to probably had "some" effect), it is not surprising that relatively more students selected "much" and "very much" (32.22% and 21.76% on average), and few students selected "little" or "very little" (11.65% and 4.83% on average). Thus, the average Likert value became significantly higher than 3. It is therefore particularly interesting to see which factors scored above average, in this case above the average of 3.55. These were the first four factors in the table, which all concern technical and scientific progress, or wars and conflicts. Far-reaching events of a structural nature, such as mass migration and population explosions did

TABLE 3.2 Results from *Youth and History* Concerning Causes of Change

Averages of Likert-values 1–5 (very little, little, some, much, very much)

How much do you think these factors have done to change life for people in history up to today?	
Technical inventions and machines.	4.39
Development of science and knowledge.	4.29
Wars and armed conflicts.	3.66
Social movements and social conflicts.	3.61
Political reforms.	3.55
Ecological crises.	3.49
Natural disasters.	3.48
Political revolutions.	3.44
Economic interests and competition.	3.43
Population explosion.	3.43
Philosophers, thinkers and learned people.	3.40
Mass migration.	3.37
Everybody.	3.26
Kings, Queens and outstanding political persons.	3.25
Founders of religions and religious leaders.	3.20

Source: This table was created by the author based on data from the database on the CD-ROM Youth and History.

not receive any high scores, nor did philosophy or religion. This confirms Shemilt's (2000) assertion that changes are perceived as being event-based. Technical and scientific progress is apparently viewed as the most important type of change, as was also the case for the associations with the Industrial Revolution.

A study of nine to 11 year old students by Barton (2008) in the United States revealed a similar picture: changes are events which cause improvements in a rational way. This is perhaps an easy conclusion to draw for technical discoveries, but the students approached by Barton (2008) also appeared to maintain these types of ideas for things such as fashion: the style "improves" and clothes are "better made" these days by using "better" materials, even if the style looks like something from the past (p. 187). Even the emancipation of African-Americans and discrimination against women rests, according to these American students, on a rationally improved insight: people discovered that African-Americans and women were simply the same as others—according to the students—and once people had understood this, then discrimination came to an end. One of the students questioned even went as far as thinking that this discovery was probably the result of medical research on African-Americans, which revealed that it was simply skin color, and nothing else, that made the difference. This greater insight immediately brought about improvements in behavior. Other students referred to the appearance of Martin Luther King: he held a speech and that led to people having better thoughts, so they stopped discriminating. Again this is a clear example of an event bringing about change. In contrast to American students, those from Northern Ireland have different ways of reasoning about change: they did not believe as strongly in progress and ascribed changes to less rational factors (Barton, 2001).

Generations

There are some data about the use of the "us-" and "them-perspective" in history education that have emerged from research into other aspects of historical reasoning. The problem is touched obliquely by VanSledright (2001) in a discussion of the ways in which a primary school teacher reacted to reading a text about the expulsion of the Cherokee Indians from their lands east of the Mississippi around 1830. The teacher showed herself emotional about the expulsion of which "we" had been guilty: "I keep saying 'we,' because I guess it is my ancestors . . . it just makes me hate my heritage almost, you know" (p. 52). She realized somewhat later that this was a story about a different time, a different place and different values and attitudes, yet the story continued to frustrate her. Her reaction was an example of transferring oneself back in time to a situation in the past, yet VanSledright wondered whether it was also an example of historical thought. Wouldn't that have required more distance, less emotion and a greater balance of judgment? A good start, for example, could have been viewing both the Cherokees and the former European Americans as "them" and studying the perspectives of both groups as objectively as possible.

The emotional reaction to the expulsion of the Cherokees is an example of a negative "us-perspective": "we" have been guilty of committing shameful deeds. More general in the United States, however, is a positive "us-perspective" belonging to the narrative about freedom, democracy and progress that dominates ideas of United States students about their own history. For instance, a ten year old talking about the first Thanksgiving: "... when we all became possible, because we all came from over there... and that's basically how we started our nation" (Barton & Levstik, 2008, p. 244). And a nine year old about the American Revolution: "before the Revolution we didn't have our rights, we weren't free" (Barton & Levstik, 2008, p. 244). This "us-perspective" seems to exist in general among students, student-teachers and teachers, irrespective of their ethnicity, origins or moment of migration of their family to the United States (Barton & Levstik, 2008; Levstik, 2000). This perspective is partially connected to what the United States represents today, for instance, when students say things like without the Revolution "we wouldn't have freedom" or "we'd still be a part of England" (Barton & Levstik, 2008, p. 244–245). However, students also use the "us-perspective" if they describe a situation in the past by itself, without referring to the situation today: "We were getting taxed and we didn't have no say about it, and we wanted some representation over in parliament, in England..." (a 10 year old student) (McKeown & Beck, 1990, p. 703); "We wanted our freedom, so people from Spain and places came over, and we could do anything we wanted to, like do our own religion" (p. 714); or about the participation in the Second World War, "We were basically just helping other countries" (Barton & Levstik, 2008, p. 250).

It seems that the negative experiences and experiences of minority groups are more likely to give rise to the use of a "them-perspective." For instance, an African-American student states that the Emancipation Proclamation "helped toward freeing the slaves" (Barton & Levtsik, 2008, p. 253) (not "us"), and the Great Depression is referred to as "what people had to go through" (p. 256) (not "what we had to go through"). Also the protesters against the Vietnam War are referred to as "them": a student likes to know "why they were against the Vietnamese people" (Barton & Levtsik, 2008, p. 257), as she is assuming as an obvious fact that the war was started to help the Vietnamese (after all, all wars involving the United States are aimed at "us" helping others), and those who opposed the war were therefore also opposing this help. Yet, also the Boston Tea Party is described from a "them" perspective: "It was when the people threw tea into the river" (McKeown & Beck, 1990, p. 703).

According to Levstik (2000) the "us-perspective" in the United States is exclusively aimed at the story of white Americans: "We" were the immigrants who drove "them" (the Indians) from their land, and during the civil rights movement African-Americans received the same rights as "us." The use of this perspective is, according to her, independent of one's own ethnicity or the moment in time at which a family migrated to the United States. The only difference of opinion appears around women's right to vote: it is the boys who say that "we" already had the right to vote, whereas girls say that "we" won the right to vote (Levstik, 2000).

British research involving students aged 14 to 18 revealed similar use of the "us-perspective" as seen in the United States. The distance in time seems to be irrelevant in this respect: "When the Roman Empire fell we were open to attacks from the barbaric Vikings and were raided frequently over the next few centuries" (Lee & Howson, 2009, p. 232). Great Britain also boasts a story of success that is easily linked in to an "us-perspective": "We're just on of the forefront, sort of leaders of democracy...And then industrially, there was the Industrial Revolution, we've been continually at the forefront of that..." (Lee & Howson, 2009, p. 237). The sense of being an island nation may also contribute to the "us-perspective," given that an island constantly had to defend itself against invaders: "...we've been on guard, from other countries and stuff..." (Lee & Howson, 2009, p. 237).

However, the "us-perspective" is regularly exchanged for a "them-perspective." For instance, great explorers are referred to as "they," whereas the British Empire is viewed as something which "we" lost (Lee & Howson, 2009). The following passage from a report by a 15 year old about the history of Great Britain is noteworthy: "...we was invaded by Normandy, a region of France, and was defeated. The next major event was the Plague. Britain was hit hard and the population went from 10 million to 2 million..." (Lee & Howson, 2009, p. 232). The change of perspective is remarkable; there is no mentioning of "us" being hard hit, or of millions of "us" that died.

Little can be concluded, based on these scarce and scattered data, about the factors that determine the use of an "us-" or a "them-perspective" and how the use of these perspectives is related to the level of historical reasoning that these students can achieve. In any event it seems apparent that students are not systematically taught to be aware of their use of either an "us-" or "them-perspective." Probably history teachers also use these two perspectives interchangeably, without being aware of doing so.

Relics

There has been little research with students that could be associated with their ability to understand the world around them as consisting of relics from different periods. According to Lowenthal (1985) it depends on a number of things whether matters are perceived from a temporal perspective: first, whether people are capable and inclined to do this, and next, on the question of how well known and commonplace things are to us (there has to be a noticeable "unusualness," but not to such an extent that this results in something being totally unrecognizable) (Lowenthal, 1985, p. 241). Something that looks "old-fashioned" (and therefore unusual) is, of course, easily perceived as belonging to a different period. There has been little research about the question of whether, and if so how, these factors play a role with students, other than the study by Levstik and Barton among primary school students, which did indeed show that something was placed in a more distant past corresponding to the extent to which it was viewed as more "old-fashioned."

A special case concerns the written relics from the past, the sources with which students are confronted regularly in modern history teaching. Do they perceive these as relics from someone from a particular historical period who wrote something down for a specific purpose? Research has shown that students have a limited inclination to locate a historical source in time, to contextualize it. For instance, Wineburg (2001) presented college students and experienced historians with a number of texts written by Abraham Lincoln. Some students read the sources without any critical distance and concluded from the contents, among other things, that Lincoln was "a racist"; others attempted to see more context around the texts (Lincoln was involved in an election campaign that he wanted to win), but still interpreted them within their reference framework from the present. In contrast to this, the historians attempted to interpret the texts against the background of historical circumstances and were aware of the distance in time between Lincoln and the present day.

In a German study among grade ten students (Körber, 2006), they were presented with a section from the hagiography of St. Boniface written by Willibald, in which he describes the saint felling a holy Germanic oak tree. When asked to explain what had happened, the students concentrated entirely on the contents of the story and even included God's help in their explanation, and in doing so adopted the author's point of view without any second thoughts. None of the students asked any questions about the author of the text, his intentions, and the circumstances under which the text had come about. In other words, the text was not seen as a relic from a distant context, but rather as a direct window on the event (Körber, 2006). Other German research also shows that students have the greatest difficulty dealing with historical documents, not only as a result of the difficult language and the insufficiency of the historical context knowledge in which events have to be placed, but also because they have trouble understanding what a historical source actually is (Langer-Plän, 2003).

These data suggest that students are probably not easily able to come to grips with the concept of "relics" as a component of historical consciousness of time, partly because of the predominant tendency toward presentism (refer to *anachronism* below).

Anachronism

To what extent are students inclined to view the past through present-day eyes or to use the past to justify the present? To what extent are they able to view the past from the perspectives of bygone times? Students seem to be able to learn not to allocate events to periods if they do not belong there, based on memorized factual knowledge. This was evident from the study by Oakden and Sturt (1922) among primary school students, which demonstrated that the students were often able to note that the Romans did not have any guns and Caesar did not wear a tall hat. This is not so much an example of historical thought, but rather of factual knowledge: you either know it or you don't.

The inclination or ability to view events from a different perspective than the one from the present ("temporal decentering") is, however, clearly a different matter. It is evident from research by Barton (2008) that the modern day perspective dominates the way in which students deal with history. For instance, they often noted that things "hadn't yet been invented," and they viewed circumstances in past events as a substitute for modern things that should have actually been there: candles were a type of replacement for light bulbs, milk used to be provided in bottles because cartons had not yet been invented. Modern day clothes are, of course, nicer than those from the past: "cause I wouldn't really like wearing that stuff, like feathers in my hat and stuff" (Barton, 2008, p. 190). British researchers have characterized this kind of attitude of students, which they also observed in their own studies, as follows: "People in the past did weird things and it does not seem to be part of everyday intuitive understanding to assume that these made sense" (Lee, Dickinson & Ashby, 1997, p. 234). A study among students aged seven to 14 confronting them with—in modern eyes—an exceptionally cruel punishment of slaves who had murdered one of their Roman masters, and with the use of trial by ordeal by the Anglo-Saxons, demonstrated that students rarely attempted to interpret Roman and Anglo-Saxon deeds from a historical context (Lee & Ashby, 2001). Explanations tended to focus on shortcomings: they didn't know any other ways to resolve the problem, they didn't have any police or courts, they didn't have any medicines, and so on.

These types of attitudes have also been found among older students, as is evident in a study from England with students aged 14 to 16, and 14 to 18 (Lee & Howson, 2009). Features of today's world are clues that can be used to translate alien matters from the past to "normal" understandable proportions: they used to speak old-fashioned Shakespearian English, now we just speak "normal" standard English; women didn't use to have any rights, and now they do; Britain's population used to be entirely white, and now it's multicultural (Lee & Howson, 2009, p. 234). The stage reached in modern life in terms of the development of democracy, science and technology appears often to be viewed as a "final goal" of history: at most, the Queen will have to give up the very last of her powers, then democracy will be "finished"; the greatest impact of the computer revolution has now passed, although the fact that technology never stands still should, of course, be taken into account (Lee & Howson, 2009, p. 237). This has led the authors to comment that presenting history at school with the aim of explaining "how we have arrived at where we are today" could in fact be promoting this kind of unhistorical presentism.

French history teachers appear to often use history as a means to justify the present day. It is apparent from research that they utilize the past to explain the "wrong" sides of it, in comparison to the present, employing sources that clearly do not satisfy modern standards of tolerance, freedom and equality (Wils, 2009). Oakden and Sturt had already pointed out in 1922 that teachers had better pay more attention to the positive details of the past, given that modern children were more inclined to notice what people in the past *didn't* have, rather than observe what they actually *did* have.

A clear modern perspective in terms of norms and values is apparent from the answers to a question in the *Youth and History* survey:

> Suppose you are a young woman/man in the fifteenth century. Your father orders you to marry John/Catherine, the son/daughter of a comparatively rich farmer in the neighboring village. You don't love or even really know your future husband/wife. What might you do, if you were living then? Consider the arguments." (Angvik & Von Borries, 1997, vol. A, p. 455)

The response options reflecting the values and conditions of the time (obeying for economic reasons, obeying because your father tells you to, obeying because disobeying your parents is against God's will, or attempting to flee to a monastery/nunnery) all ended up below the average value of 3 on the Likert scale (2.18, 2.34, 2.23 and 2.02 respectively on average), while the two most recent options (refusing because it is immoral to marry someone you don't love, and refusing because it is everyone's right to marry for love) clearly resulted in scores above 3 (3.81 and 3.90) on average (Angvik & Von Borries, 1997). There has been some criticism of the question, as it was posed in the form of "what would you do if. . . ," and not as "what would a girl or boy in that time have done. . . ." This supposedly would lead students to answering from their own, that is modern day, perspective. It is impossible to ascertain from the large Youth and History survey what deliberations the students went through when answering the question. However, a similar type of study conducted in Germany did demonstrate that a modern day perspective was indeed applied to the past (Körber, 1998). This is not surprising given the results from the American, British and French studies discussed previously.

Contingency

One of the aspects associated with the key concept of contingency is the fact that people from later times have knowledge of the outcome of events, unavailable to those who were present at the time events took place in the past. To what extent is a known outcome of events also viewed as "logical" and "unavoidable," and to what extent is the possibility of another outcome taken into consideration? Research by the psychologist Fischhoff (1975) revealed some important details about this question. He presented university students with a number of situations that included four possible outcomes for a historical event. One of these involved a nineteenth century fight between the British and some Gurkhas with the potential outcome being either a victory for the British or the Gurkhas, or an indeterminate outcome with or without a peace treaty. He created five groups: one group was presented with a version of the story with none of the outcomes as "actually happened," and four groups were presented with versions of the story where a single sentence was included at the end of the description indicating one of the four options as "the real outcome." Apart from that single final sentence, the stories were identical. The groups were asked to indicate which outcome they believed to be most likely, based on the description of the event (the terrain, the relative strength of both armies, the battle experience of both armies, command structure, etc.) and to indicate which factors they believed to have played the most important role in this. The result was that the groups who had been presented with a certain outcome systematically indicated a higher likelihood of that event taking place. They also placed a greater weight on the factors that had increased the likelihood of this outcome. Further research demonstrated that they were not aware of the influence of their knowledge of the outcome on their assessments. Fischhoff (1975) believed these results were highly significant for the study of history:

> It might be asked whether this failure to empathize with ourselves in a more ignorant state is not paralleled by a failure to empathize with outcome-ignorant others. How well people manage to reconstruct the perceptions that others had before the occurrence of some event is a crucial question for historians, and indeed for all human understanding. (p. 295)

Fischhoff's first study into "hindsight bias" was followed by a series of studies into problems that people experience in ignoring information available to them and not to others. This research was summarized by Royzman, Cassidy and Baron (2003) under the collective term "epistemic egocentrism," which they defined as

> a specific failure of perspective taking that results from a difficulty in . . . setting aside the relevant privileged information (knowledge) that one knows to be unattainable to the other party, with a result that one's prediction of another's perspective becomes skewed toward one's own privileged viewpoint. (p. 38)

The research discussed by them from the period 1975 through 2000 demonstrated, according to these three writers, that this phenomenon occurs both in children and adults, although to a greater extent in children. This observation leads to the surmise that aspects related to the contingency concept as a component of historical time awareness might pose considerable problems to students.

A study that seemed, at first glance, to contradict this was the one undertaken by Hartmann and Hasselhorn (2008). They presented students with a (fictitious) conversation between two young men in 1930 in the Weimar Republic. In this conversation the young men were discussing who they were going to vote for. They were complaining about the economic and international/political situation in Germany and were arguing that the country needed a strong leader. The question was whether Hannes (one of the young men) would vote for the NSDAP. The students had to indicate on a 0–3 Likert scale to what extent nine considerations applied to Hannes' situation. Of the nine, three considerations were included that presented a knowledge of the outcome of events (e.g., Hannes won't vote for the NSDAP, as what the Nazis did to the world can't be condoned, or the NSDAP's slogans are easy to unmask; it is clear that this will lead to war) and three were clearly presented within the context of that period (e.g., Hannes has no experience of democracy; he does not know how risky it is to vote for the NSDAP and will therefore vote for that party, or Hitler represented a strong leader to him; he will not consider how dangerous it is to vote for him). The students involved in this study were on average sixteen years old and were attending the *Gymnasium*, the highest level of secondary education in Germany. The Weimar Republic had been extensively discussed during the academic year in which the study took place. The deliberations corresponding to the prevailing context of the nineteen thirties were chosen more often by these students than those based on knowledge of the outcome. It could be concluded from this that they were indeed able to select the perspective relevant to that time period. However, is this true? After all they *knew* that many German voters had indeed voted for Hitler around 1930 on the basis of the deliberations such as the two discussed above. The fact that the Weimar Republic had been extensively discussed in their history lessons suggests that this was highly likely. All three deliberations based on knowledge of the outcomes led to the choice of *not* voting for the NSDAP, while all three deliberations without outcome knowledge led to the choice of indeed voting for the NSDAP. The knowledge about the outcome of the elections is therefore also a form of hindsight in this example. Students knew that many voters in the nineteen thirties voted for the NSDAP based on the considerations of Hannes without outcome knowledge, and therefore it is logical that many students chose for these "contextual" considerations (Hartmann & Hasselhorn, 2008). In this form this study says little about their capacity to place themselves in the situation of being a voter in the Weimar Republic in 1930. It would perhaps have been more enlightening to also include some considerations without outcome knowledge (so "contextual" ones) that would have led to *not* voting for Hitler.

I am not aware of any research that sheds light on the extent to which students are inclined to take unintended consequences into account in history. With respect to the logical and straight line from the past to the present, the impression based on research by Barton (2008) and by Lee and Howson (2009) previously discussed, is that students normally reason in terms of a past that is entirely focused on the present day as an unavoidable outcome. The belief in progress is particularly present among American students, who attribute the majority of developments to science and technology. Students seem in general not to consider historical events as having unpredictable or uncertain outcomes and presumably have little consideration for the possibility that the present could have looked significantly different if events from the past had had different outcomes. Barton and Levstik (2008) note that "their progressive view of history excluded the consideration of alternatives" (p. 248). It is therefore not likely that they view themselves as having been determined, more or less, by fortuitous historical developments.

Conclusion

The observations discussed in the earlier sections of this chapter regarding young students and their ability to deal with chronology also seem to apply broadly speaking to older students and even adults: knowledge of the past exists in images and contexts (separate islands of time), rather than in the form of a chronological line, and mathematical chronology is not easily accessible. Given the fact that these insights are partly based on psychological research with adults, it is not surprising that they should also apply to older students beyond primary school as is evidenced by results from *Youth and History* (Angvik & von Borries, 1997).

Associations with diverse periods seem to materialize particularly in the form of concrete and visible events and less so in the form of background factors. It is unclear to what extent there is a differentiated image of a historical past. Perhaps stereotypical images about "back then" (whatever period in the past) as full of repression and strife are very much alive for many students. The associations with the Middle Ages, the Colonial Age and the time of industrialization in the *Youth and History* (Angvik & von Borries, 1997) survey may be interpreted in this way. This may indicate the presence, to a certain extent, of a "mythical" past.

The way in which students deal with the relationship between continuity and change and the associated periodization has not been explored. Research regarding the insights into changes has indicated that these are perceived as "events" rather than processes or developments. A great deal of weight is attached to discoveries and scientific progress in this context. Changes are viewed as being logical and rational: there was a problem, someone thought of something clever, and then it changed (improved). Whether this is a specific American perspective needs to be explored in further detail.

The "us-perspective" and "them-perspective" are used in different ways for reasons that are not entirely clear. It is apparent that the "us-perspective" is not systematically avoided when talking about the historical past.

Presumably this is also the case for history teachers. Research suggests that a strong bond is felt between the past and the present and that the past is seen as a precursor to the present, which logically flows from it, as is evident, for instance, in the observations by the American students regarding the significance of developments, such as the American Revolution, the Civil War, and the Civil Rights movement.

Students' perspectives on material relics have not been investigated, although it is clear that something that looks "old-fashioned" is more easily placed in the past. In terms of written sources it is clear that students have great difficulty with viewing these as products that have been created within a given historical context with particular aims.

Avoiding anachronisms, insofar as these are based on knowledge of factual circumstances in a given time period, is no more difficult than learning about historical facts in general. A different matter is assessing and understanding situations according to the standards of the time. Research indicates that the opposite is usually done by many students; present day standards and a significant level of presentism appear to dominate their approach to history. The use of the past as a "bad example" by teachers seems to unintentionally promote this.

There has been no direct study into the extent to which students use history to justify current viewpoints; we can surmise that this may occur, given the experience of a meta-historical identity common to people in the past and the present by many students, as is apparent from their use of the "us-perspective."

Psychological research has indicated that it is particularly difficult or almost impossible to ignore outcome knowledge. It is anticipated therefore that this knowledge will play a significant role for students when studying history. Presumably they are not aware of this influence and the prejudice that may follow from it as a consequence. An awareness of the capriciousness and unpredictability of the course of history, often with unintended consequences of human behavior, also seems to not be well developed as standard; rather students seem to see a logical and straightforward development characterized by progress. History teaching may perhaps unwittingly contribute to this by trying to explain "how we arrived at where we are today."

There has been no direct research into the extent to which students view themselves and their own thoughts, religious beliefs and convictions as a product of a contingent historical process, of which the outcome could have been completely different; however it is obvious from the discussion above that there is little evidence of this.

An overview is provided in Table 3.3 of the research agenda formulated at the end of Chapter 2. It shows to what extent the previously formulated research questions may be answered by existing research. It should be noted here that there is little differentiation in terms of various age groups, and therefore only general impressions have been provided for what "students" normally do. Further research is required to be able to provide a more nuanced image.

TABLE 3.3 Answers to Research Questions Concerning Historical Consciousness of Time Based on Existing Research

Question	Answer as far as can be given based on existing research
1. To what extent is an awareness of daily, cyclical time, social time and a "mythical" time present in students and to what extent is an orientation towards a long-term, linear temporal awareness a deviation from this in their experience?	*No data.*
2. Are references to time that make use of calendars and years more difficult to remember than references that make use of an inherent qualification of a moment in time?	Based on psychological research: this seems to be the case. Scarce research in teaching and learning history confirms this surmise.
3. To what extent do students tend to perceive time periods as something actually existing, as given entities corresponding to an historical reality?	*No data.*
4. Are imaginative-associative names of time periods (like "time of the knights," "age of steam") more easily remembered than neutral references (like 1100–1500, 19th century)? If so, are imaginative-associative temporal references supportive in creating an historical reference framework?	Based on psychological research: this seems to be the case. Scarce research in teaching and learning history confirms this surmise. See question 2.
5. To what extent do students tend to underestimate temporal differences within a time period and overestimate temporal differences between periods? To what extent do students tend to overestimate differences between time periods and underestimate correspondences between periods?	*No data.*
6. To what extent do students tend to and are able to see that noting discontinuities rests on interpretations, and to what extent are they able to distinguish between processes of change at different levels?	Data about changes as "events" and "rational progress."
7. To what extent do students distinguish between an historical and a social past?	*No data.* See question 1.

(continued)

TABLE 3.3 Answers to Research Questions Concerning Historical Consciousness of Time Based on Existing Research (continued)

Question	Answer as far as can be given based on existing research
8. To what extent and under what circumstances do students tend to develop an "us-" or "them-relationship" with people from the past?	Data about the use of the "us-perspective" in the United States and Great Britain. It is still unclear what influences the use of either the "us-" or the "them-perspective" and how this is influenced by (history) education.
9. To what extent do students tend and are they able to perceive the present world as comprising relics originating from different time periods in the past?	Data about students' problems with contextualization suggest that this is a learning problem.
10. To what extent do students tend and are they able to perceive the "non-contemporaneity of the contemporaneous" in a cultural environment in the past?	See question 9: Probably we have to do here with a larger learning problem connected to that question.
11. To what extent do teachers and students in history education tend to make use of or avoid using phrases suggesting that things were "not yet" or "already" the case in the past?	Data about the use of the past as "the wrong example": probably the use of words like "already," "still" and "not yet" is not consciously avoided.
12. To what extent do teachers and students tend to use the past to justify certain claims in the present?	Data about history as the story of "how we became what we are now." Probably history is being used as justification for the present situation.
13. To what extent do students tend to and are able to explain developments in the past as unintended consequences of human actions or a coincidence of circumstances?	Data about the supposedly rational character of changes and the estimated probability of outcomes such as are known to us suggest: probably there is not much consideration of unintended consequences.
14. To what extent do features of "epistemic egocentrism" apply when confronted with people from the past?	Data about the role of outcome knowledge suggest that these play a role in history education.
15. To what extent do students tend to view the path from the past to the present as a direct and logical line of development?	Data about the past as prehistory of the present suggest: probably people tend to distinguish this kind of direct lines. See question 13.
16. To what extent do students tend to and are able to relativize their present situation, insights, (religious) beliefs and convictions as historical phenomena?	Given the data such as mentioned for question 13 and 15, it seems unlikely that students have this kind of inclination.

Conclusions and Consequences for Future Research

Thinking about time belongs to the category of things that are relatively difficult for humans to achieve. This is due to the fact that it requires "temporal decentering": leaving one's own perspectives of the present and transferring oneself into the past or future. Transferring one's thoughts in time is more difficult than transferring one's thoughts to somewhere else in space, as space is observable in concrete terms, whereas time is not. This is why humans use spatial metaphors in any language—as far as is known—to reflect distances in time.

A historical consciousness of time relating to many hundreds of years in a linear image of consecutive periods does not arise by itself. It can only be the result of conscious study, as has primarily been the case in Western culture. It is therefore logical that the development of a historical consciousness of time comes about through a learning process, not a developmental process. Human memory is, in principle, poorly equipped for the storage of information in a linear pattern of singular events. This is probably because information like this is rarely required for daily functioning, whereas information about cyclical, recurrent events is more important. If we nevertheless intend to promote a historical consciousness of time, then this process may be facilitated by linking it to the way in which people store information about time in their autobiographical memory. This is done by creating a kind of time islands: contexts and locations in memory. Later on these locations may be coordinated by means of a mathematical chronological system, but this doesn't always necessarily happen. There are reasons to believe that the evidence from psychological research about autobiographical memory also applies to learning historical information. If that is correct, then it would follow that no precedence should be allocated to numbered years and timelines, but that efforts should firstly be made to building associative contexts, which may subsequently be ordered into years and timelines.

A historical consciousness of time not only reflects the image of a long-term, linear course of time with successive appearances of different periods; it also concerns the relationship people have with that image. Within the existing research there is some evidence regarding certain aspects of historical consciousness of time, while there are still large gaps of knowledge about other points.

For instance, it may be stated with some certainty that students tend to perceive changes as events, often relating this with rational decisions from individuals, which led to progress. The relationships between change and continuity, change processes of various natures and tempos, and structural changes and developments that have led to setbacks receive little attention in this respect. Furthermore, it is apparent that students tend to perceive the past as a precursor to the present, taking the present as the ultimate goal at which virtually all past developments have been directly aimed. The fact that history education frequently aims to use the past to explain how the world has become what it is today probably contributes to this. The present is generally used by students as the norm by which the past is to be judged and evaluated. They view events from the perspective of their own time (presentism). History teaching may also contribute to this through the use of the past as a "bad example." Research has shown that knowledge of the outcome of a historical development leads to that outcome being perceived as the most likely outcome and that it is very difficult to disregard the knowledge of the outcome. This also contributes to the fact that historical developments are perceived as unavoidable and that there is little attention for other potential turns that those developments could have taken in the past. Finally, it seems clear that students have great difficulty viewing cultural relics, particularly written sources, as products of a particular time and particular context and interpreting them from this perspective. Sources are often seen as a direct window on the past.

All of these assertions indicate that the development of a historical consciousness of time is a difficult task. Additional research should be able to provide further qualifications and revisions to the data collected here. The results of this could be used to develop new educational strategies. However, there is a greater urgency for research into those aspects of historical consciousness of time about which there is little or no clarity. As is apparent from Table 3.3, this concerns research questions 1 through 5 and 7 in particular. These concern three topics: the question of whether a daily, social and mythical awareness of time plays a role for students in contrast to a historical consciousness of time (research questions 1 and 7); the question of how students deal with the interpretive character of periodization (research questions 3 and 5); and the question of whether the location theory of memory, the theory of associative contexts, facilitates learning historical information about time, and whether it is consequently correct that the learning of associative contexts needs to be prioritized above learning numbered years and timelines (research questions 2 and 4). The question of the use of the "us-perspective" and "them-perspective" deserves attention among the remaining research questions for which there are at least some rudimentary answers: under which conditions are these two perspectives chosen? This depends, to some extent, on the question of whether students tend to perceive their own situation as being determined by history, and therefore as coincidental and variable. The more coincidental and variable the current position is considered to be, the greater the space for a "them-perspective" and less of a tendency there will be to justify the current situation on the basis of the past. This appears to be a logical assumption.

The following chapter will report the results of an empirical study into research questions 2 and 4. These two questions were selected, as the psychological and educational research have created a relatively clear and testable hypothesis in this respect.

4

Empirical Research into Students' Orientation in Historical Time

The psychological and educational research described in the previous chapter suggests that awareness of time rests first on associative contexts, and only subsequently on chronological dates. Applied to teaching and learning history, this would mean that students would—in order to develop a historical reference framework—firstly make use of associative contexts, and only secondarily focus on numbered years and timelines. The hypothesis being tested in this chapter with the aid of empirical research has been based on this, and reads as follows:

> References to time utilizing numbered years and numbered centuries are more difficult to learn and to retain than references to time utilizing imaginative-associative contexts; imaginative-associative contexts have a positive effect on the creation of a historical reference framework.

Images of Time, pages 143–182

A historical reference framework is defined as: a (temporal) framework enabling students to locate not only known events and phenomena correctly in time, but also events and phenomena that are new to the student.

The development of a historical reference framework appears, in practice of history education, not to be an easy task for students. In order to be able to effectively manage a framework such as this, students have to:

- know which periods can be distinguished and in which order they need to be placed;
- be able to name events and phenomena from each period;
- be aware of and be able to recognize characteristic aspects for each particular period, so that "new" events and phenomena can be placed within the correct time period.

Research has shown that students have difficulty learning to do this. In the context of the project *Usable Historical Pasts*, a number of approximately 14 year old British students were asked whether they were able to describe—in broad terms—the events that had occurred in Great Britain over the last 2000 years. A significant proportion of the students refused to carry out the assignment because they had no idea how to approach a task like this. Other students named random events and phenomena without any connection or sequence. Only a small minority managed to produce a coherent summary on paper (Howson, 2007). German research has also demonstrated that the extent to which students succeed in mastering a general outline of major developments in history is disappointing (Popp, 2006, 2009). The lack of a reasonable overview of periods and their characteristic aspects will probably lead to a total absence of what is called a "sense of period," an ability to recognize periods, which is created by diverse kinds of knowledge, such as: what the world looked like during certain periods (buildings, clothes, tools); what people's opinions, convictions and thoughts were; issues about government and authority; wars, social inequalities, and so on (Dawson, 2009). Before arriving at the stage of a developed "sense of period," one needs first to acquire some basic knowledge: When did a certain period occur; how does this period fit in with other periods; who lived during that period and what events occurred?

The experiment described in this chapter concerns the question of what the optimum way is for students to form an image about a particular piece of the historical past that provides an answer to the latter types of questions. An experiment has been conducted to find out whether students are best able to distinguish between various periods, within a unit of historical subject matter that is unknown to them, by means of numbered years and centuries or by means of imaginative-associative contexts, and whether they are better able to locate events in time using numbered years and centuries or imaginative-associative contexts. This study does therefore not involve a full-scale contextualization of historical phenomena and events, which would also involve integral associations between known contexts and new events and phenomena. Nor does this study investigate the manner in which students develop a sense of time periods.

Research Design

It was apparent from the previous chapters that the chronological component of a historical reference framework is relatively foreign to the way in which people prefer to orientate themselves in time. However, within history education it is often assumed that mastery of chronology should be an important foundation to studying a unit of subject matter. In one of the first history lessons, students often have to create timelines, dealing with years before and after the birth of Christ, and distinguishing between centuries. In history textbooks, chapters often begin with a timeline indicating the most important events. Numbered years are often used as anchor points when studying subject matter after it has been discussed in class.

An experiment was designed to investigate the role played by chronology (years, centuries, timelines, etc.) and associative contexts in the development of a historical reference framework. The experiment involved two parallel groups of students who were taught the same topic using two difference teaching methods. These two teaching methods can be described as follows:

- The chronological approach is first orientated on numbered years and centuries. The placing of events in the correct sequence and in the correct century plays a principal role in this approach. Years and centuries form the backbone to order and present the study material. In this instance connections between new details and the context that has already been learned are made via numbered years and centuries, which have to turn into fixed memory locations.
- The imaginative-associative approach is aimed first at creating certain locations in memory that are linked to images that can facilitate associations. Narratives about historical situations are presented in order to create images, which can serve as a focal point for associations. No distinctions are made between different times through the use of numerals—as, for instance, is the case for cen-

> turies—rather, characteristic names are used for eras. These names reflect an important aspect that is characteristic of a time period. The question is whether this approach also creates memory locations that can be used to link in new details, that is, events that are not covered by the name of a particular time period.

Two teaching units were developed on the basis of these descriptions, presenting the same historical content extending over a longer period of time, namely some hundreds of years. An experiment based on a shorter time period would not sufficiently challenge students' capacity to orientate themselves in historical time. Both contents were presented to two parallel groups of students at secondary schools who were of the same age and same level of education: intervention X_1 and intervention X_2 (see Figure 4.1). The effect of the interventions was determined through a post-test (O_2), consisting of the task of locating 25 events in historical time. A pre-test (O_1) was used to investigate the students' initial status. For the pre-test, the students were asked to locate 25 events correctly in time, the same ones as were used for the post-test. The pre-test and post-test were the same for both experimental groups (groups A and B).

This study involved a quasi-experimental approach, as neither group of students (group A and group B) represented a random sample of students. The lessons were carried out in existing school classes, which meant that small differences between the groups could not be ruled out. Further details are provided below regarding these potential differences between the groups. The quasi-experimental method is a common approach within

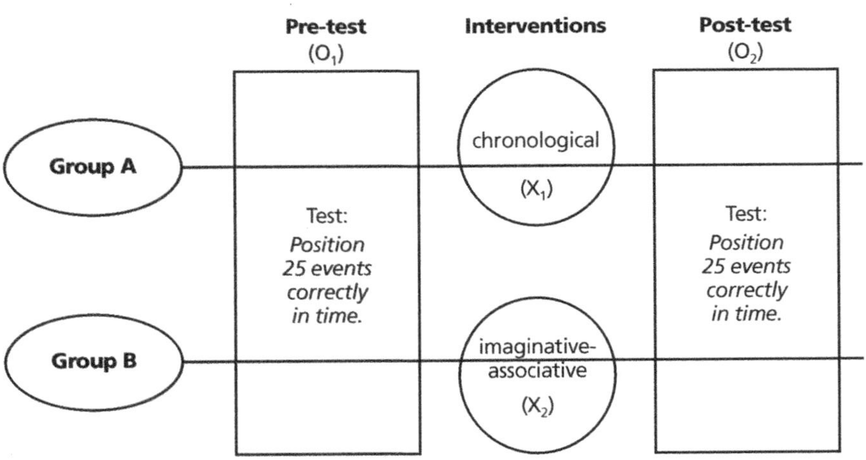

Figure 4.1

educational research, as laboratory situations involving students are difficult to create and are undesirable for ethical reasons (Cohen, Manion, & Morrison, 2007). The design discussed here is an example of a quasi-experimental design making use of a pre-test and a post-test and a comparison between two parallel groups (Shadish, Cook, & Campbell, 2002). The difference here compared to standard quasi-experiments is that an intervention was employed in both groups to compare the effects between them. The more usual approach is to apply an intervention to one group and give the other group the normal treatment without any intervention (the "control group"); this allows the clearest assessment of the effect of the intervention. However, it was not possible in this instance to omit all interventions in one group, as will be clarified later. This means that a perfect comparison of the effects in both groups becomes a crucial aspect of the study design.

A problem for any quasi-experimental study is that different outcomes on the post-test may potentially not be (wholly) attributable to the effect of the intervention, but rather to differences between the groups existing previously. Therefore, in conducting this study, a great deal of attention was paid to the composition of the groups and measuring any potential differences between the groups.

Composition of the Groups of Participants

Groups A and B were drawn from classes from four different secondary schools from different regions in The Netherlands (Leiden, Amsterdam, Alkmaar and Leeuwarden), creating two groups of 135 and 132 students, respectively. The size and composition of the groups enhanced the potential for mutual comparability and the representativeness of the sample in comparison to the total student population (of the relevant age and school level) in the Netherlands (external validity) (Cohen et al., 2007). Group A and Group B consisted throughout of parallel classes of students from the four schools: four classes of grade eight students in Leiden, two classes of grade eight students in Amsterdam, two classes of grade eight students in Leeuwarden, and four of classes grade nine students in Alkmaar. This allowed Groups A and B to be formed from students out of six classes from four different schools. The parallel classes were comparable throughout in terms of their composition; they were "random" in the sense that the school classes had not been separated on the grounds of different student characteristics, but only due to the fact that the number of students in a given age group was too large for a single class. Three out of the six pairs of classes were normally taught history by the same teacher. During the experiment, each pair of parallel classes were taught by the same teacher.

Allocation of the classes to either group A or group B was random. This meant that neither the researcher (the present author), nor the classroom teachers nor the school influenced the allocation. The intervention was carried out by guest teachers specially chosen for the task, who were responsible for applying approach A or B to a particular class. This choice was again random, given that the visiting guest teachers had neither visited nor seen the classes prior to applying the intervention. The guest teachers did not consult with the researcher about which option to apply for which class. The classes were not informed of the differences between the two teaching methods with which they were going to be presented (they were not told what the other parallel group had been presented with and therefore had no reason to pay any specific attention to particular issues). These "screening" measures probably had a positive effect on the validity of the experiment (Shadish et al., 2002).

The composition of both groups created using this method is shown in Table 4.1, and includes the educational level (intermediate secondary education or advanced secondary education), numbers of male and female students, numbers of native and non-native Dutch children, and their ages. Students born outside the Netherlands or those with one or both parents having been born outside the Netherlands were categorized as non-native Dutch children.

As is apparent from Table 4.1 there were some more advanced level students in group A compared to group B. The number of male and female students in both groups was virtually identical. The number of native and non-native Dutch students was virtually identical, as was the age distribution (as shown in the table there were a number of students who did not provide their age).

The students were asked whether they enjoyed history as a subject (yes = 2, a little = 1, no = 0). Their teachers were asked to provide the students' last report grade. The averages of these figures are shown in Table 4.2.

The differences were small, however, insofar as a difference can be discerned this was to advantage of group A. The results of a *t*-test revealed, however, that the differences between the groups were not statistically significant (appreciation for history: $t = 0.308$; $df = 256$; $p = 0.758$; report grade: $t = 1.290$; $df = 239.04$; $p = 0.198$). The standard deviation for the report grades was greater in group B than in group A, meaning that the differences between higher and lower grades were greater in this group than in the other group.

TABLE 4.1 Features of Students Participating in the Experiment

Group		Number	Percentage
Level			
A	Intermediate	62	45.9
	Advanced	73	54.1
	Total	135	100.0
B	Intermediate	69	52.3
	Advanced	63	47.7
	Total	132	100.0
Sex			
A	Male	67	49.6
	Female	68	50.4
	Total	135	100.0
B	Male	56	42.4
	Female	76	57.6
	Total	132	100.0
Natives and Non-natives			
A	Natives	118	87.4
	Non-natives	17	12.6
	Total	135	100.0
B	Natives	115	87.1
	Non-natives	17	12.9
	Total	132	100.0
Age			
A	13	23	17.7
	14	72	55.4
	15	30	23.1
	16	5	3.8
	Total	130	100.0
B	13	23	17.7
	14	72	55.4
	15	30	23.1
	16	5	3.8
	Total	130	100.0

Note: Some students failed to fill out their age on the questionnaire. The age distribution among those that did was—surprisingly and coincidentally—exactly identical in the two groups.

TABLE 4.2 Appreciation of History and Report Grades

	Group	N	Average	Std. deviation
Likes history (0–2)	A	133	1.19	0.947
	B	125	1.15	0.925
Report grade for history (1–10)	A	135	6.696	0.9641
	B	127	6.520	1.2270

Note: Some students failed to answer the question about liking history and not in all cases it was possible to retrieve reliable report grades.

Description of the Interventions

The interventions in group A and group B consisted of a single lesson (50 minutes) in which students were presented with a topic that they were expected to know little or nothing about. The "history of Aruba from 1400 to the present" was chosen as the topic for the following reasons:

- It is a topic that is not taught as part of the usual curriculum and about which little is known and about which there are probably few general impressions outside of education. It was anticipated that the baseline condition in terms of the knowledge of this topic would be zero or virtually zero among the students. The "disruptive" influence of historical knowledge acquired through the usual teaching was potentially minimal for this topic.
- It is a topic that stretches across six centuries, so orientation in time plays an important role.
- It is a topic that, relatively speaking, is not very complicated or "extensive," so a reasonably "complete" overview could be provided in an concise lesson format.
- It is a topic that involves an autonomous territory within the Kingdom of the Netherlands, meaning that a reasonable amount of interest in the topic could be expected from the students in the Netherlands—despite the fact that it was largely unknown to them—meaning that any reference framework created would not be unduly and negatively affected by disinterest.

A whole lesson of 50 minutes was chosen; this was to be presented by an experienced history teacher presenting the history of Aruba via either a chronological or imaginative-associative approach to groups of students. This choice was made to make the lesson appear as much as possible like a "normal history lesson." This increased the ecological validity of the experiment, meaning that the experimental situation did not deviate too much from the usual educational environment, which could negatively affect the assessed learning outcomes (Cohen et al., 2007; Bryman, 2008). The option of presenting only written information would have excluded any risk of the teacher affecting the outcome of the learning process in potentially different ways; but the confrontation with the subject matter, and therefore with both teaching methods, would in that case have been a lot less intensive. In order to avoid differences between teachers and their relationships with classes from playing a role, two guest teachers were recruited who were specifically asked to cooperate with the experiment. The guest teachers had considerable experience as history teachers, but were not teachers at the schools at which the experiment was conducted and were therefore unknown to the participating students. One and the same teacher was used for a lesson presented to two parallel groups according to the two different approaches. Two teachers took part in this: one of them taught at the schools in Leiden and Leeuwarden, the other in Amsterdam. Both guest teachers were employed as teachers in Alkmaar.

A PowerPoint presentation was developed for both lessons (see Appendix A) and a protocol was given for the contents of the lessons (see Appendix B). For the historical contents of the lessons and tests, use was made of Hartog (1980, 1993), Daal (1992) and Dalhuisen, Donk, Hoefte, & Steegh (2009). Sound recordings were taken of the lessons. These demonstrated that there were no significant differences between the pre-determined lesson contents and the lessons as they were actually carried out. It became apparent from interviews with the teachers after the experimental lessons that—other than the difference in the teaching method—they thought the lessons had been equivalent: neither lesson was better, more enjoyable, or more agreeable than the other.

For the *imaginative-associative* approach each time period was introduced by a name (Time of the Indians, Gold, Horses and Pirates, etc.) and then the subject matter was introduced by means of a small story to provide an illustration for the name of the period. This narrative context was intended to act as a memory location with which different items could be linked up. Subsequently, a number of other events and occurrences were presented for the same time period. For the *chronological* approach each time period was presented as a numbered century, and a number of anchor points were discussed using numbered years for events in order to place this time period on a chronological line. The numbers for the centuries and the years were intended to act as memory locations to allow a number of events to be linked up. Subsequently, a number of other events and occurrences were presented for the same time period. These latter events and occurrences were the same as those presented in the imaginative-associative approach. Figure 4.2 shows

Chronological approach	**Imaginative-associative approach**
fifteenth century	Time of the Indians
sixteenth century	Time of the Useless Island
seventeenth century	Time of Horses and Pirates
eighteenth century	Time of Gold
nineteenth century	Time of Oil
twentieth century	Time of Tourism

Figure 4.2

the time periods used for both approaches. An explanation for the meaning of the associative names for approach B can be found in the text for the lesson protocol belonging to approach B (Appendix B).

The use of numbered centuries and years alone as anchor points for the chronological approach is to some extent artificial in comparison to standard history teaching, because usually period names, such as Renaissance or Enlightenment, are used alongside numbered years and centuries that will allow associations to be made. However, in the context of this study it made sense to draw a clear distinction between purely chronological and purely imaginative-associative aids in order to distinguish between the time periods. After all, the question was which aids would best support orientation in time: the chronological or imaginative-associative.

Further explanation is provided below regarding the difference between the two approaches using the example of the nineteenth century and Time of Gold.

The title slide for a time period (Figure 4.3) in the *chronological approach* consisted of the indication of the concerned century in large and clear font, for example, the 19th century. The numeral nineteen was included on all slides in the top right-hand corner throughout the presentation of this century. The students were told that this time period ran from 1800 to 1900. In the *imaginative-associative approach* an image was displayed in the center of the slide representing the time period, in this instance a heap of gold bars and gold coins. There was an explanation that this was the "Time of Gold" and that this period ran from 1800 to 1900. This is the only time within the imaginative-associative approach that use was made of (whole figures for) years: these were always included at the start of the time period. The image representing the time period remained visible in the top right-hand corner of the slides during the discussion of the time period.

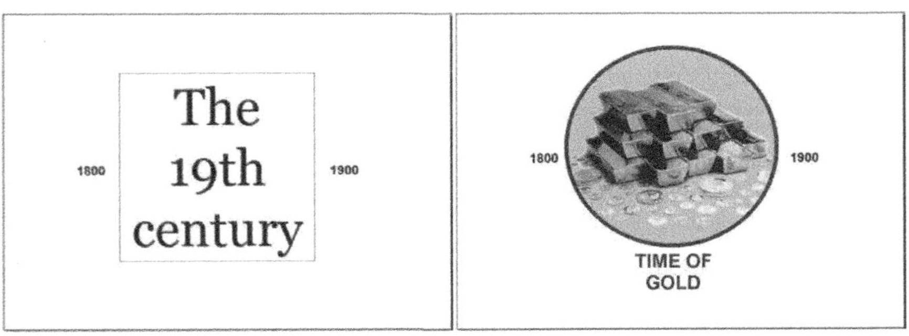

Figure 4.3

In the chronological approach the time period was introduced through a number of chronological details on the second slide (Figure 4.4): Aruba had been conquered by the English during the Napoleonic period, but was returned at the end of this period to the Dutch as a colony. The Dutch government ordered a survey to be carried out to see what could be discovered on its new colony. The result was that there was very little to discover; the island barely managed to produce enough to support its own inhabitants. At the bottom of the slide, a timeline is displayed showing the position of the nineteenth century; this remained on the screen during the discussion of this century (the nineteenth century was indicated on the timeline with a contrasting color, which cannot be seen on the black and white copy of the slide). Parallel to this, two slides were used for the imaginative-associative approach to tell a story. A shepherd boy is sent into the mountains without shoes to herd the sheep and by accident finds a lump of gold providing more than enough money to buy him a pair of shoes. As a result of this discovery a multitude of gold diggers arrive on the island, but they discover very little gold. The story presented on these slides is supported with little "speech bubbles" with people's expressions and thoughts.

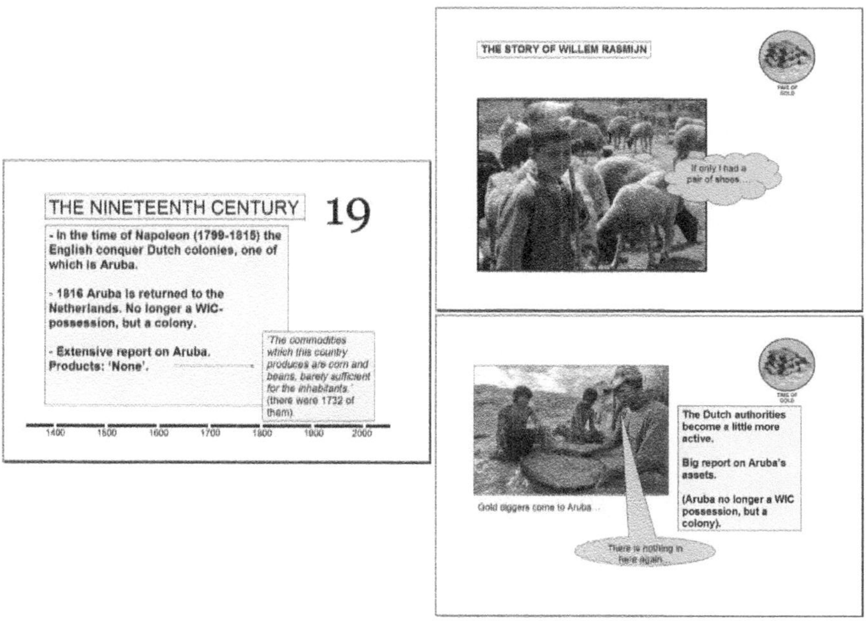

Figure 4.4

Figure 4.5

For both the chronological and imaginative-associative approaches, information was now presented about the gold mining (Figure 4.5): for the chronological approach this was achieved with the inclusion of references to two particular years, and for the imaginative-associative approach the information was more global. The images presented were the same, although for the imaginative-associative method the images could be larger as less chronological information needed to be presented.

The balance of the gold production was produced (Figure 4.6). The information presented here was exactly the same for both approaches; the only differences were the timeline, century number or image presenting the time period.

The second aspect of this time period was introduced (Figure 4.7): phosphate production, which in economic terms has been much more important than gold production, as it has produced greater profits. For the chronological approach a few numbered years were included once again around the discovery and extraction of phosphate. For the imaginative-associative approach a link was made to the gold story (somewhat

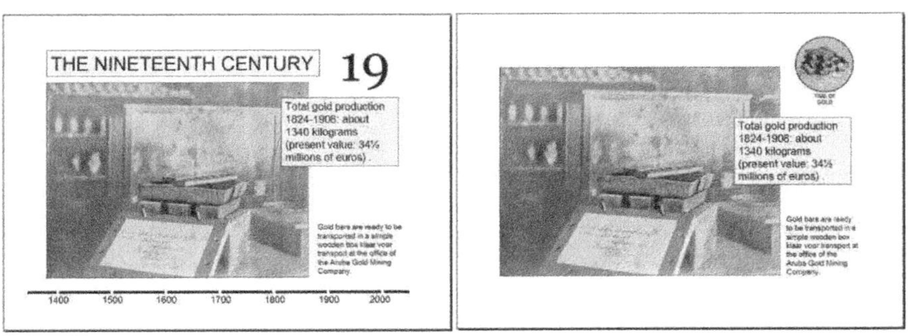

Figure 4.6

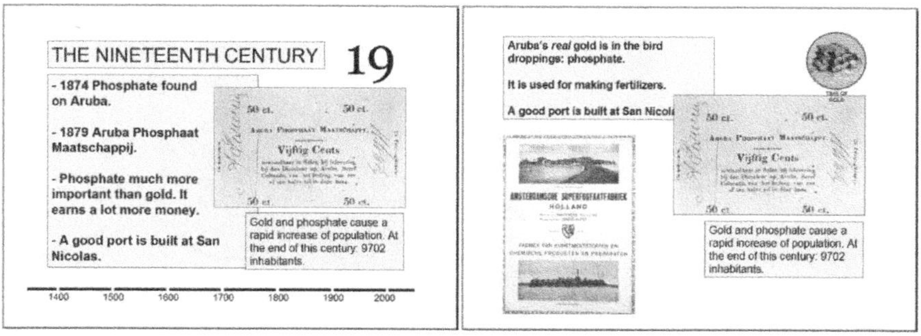

Figure 4.7

expressively) by saying: "the *real* gold on Aruba was in bird droppings." An image of the bank note issued by the Aruba Phosphate Company and information about the result of phosphate production were identical. For the imaginative-associative approach there was space for an extra image (the Amsterdam phosphate factory) instead of the chronological information on the corresponding slide.

In the next slides (Figure 4.8) both approaches were identical: a harbor was built as a result of the phosphate production, which was of importance as oil tankers were able to moor within it in the following time period.

The last slide (Figure 4.9) for each of the time periods was one that included the balance: what were the most important things in the nineteenth century or: what does the Time of Gold bring to mind? For the latter, associative thinking was emphasized by placing the image representing the time period in the center surrounded by associative "bubbles" where students could fill in their associations (this was done orally during the lesson).

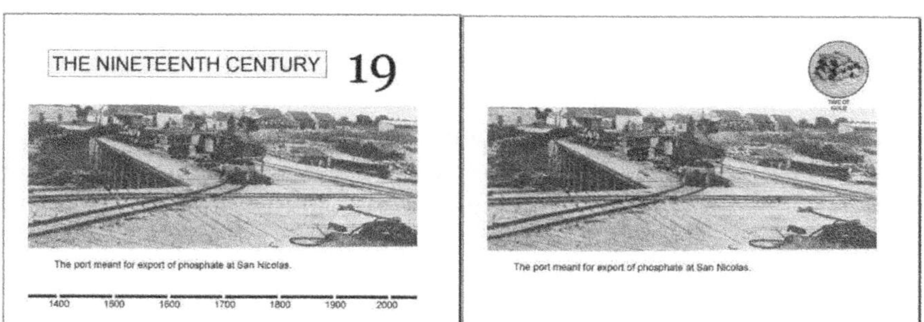

Figure 4.8

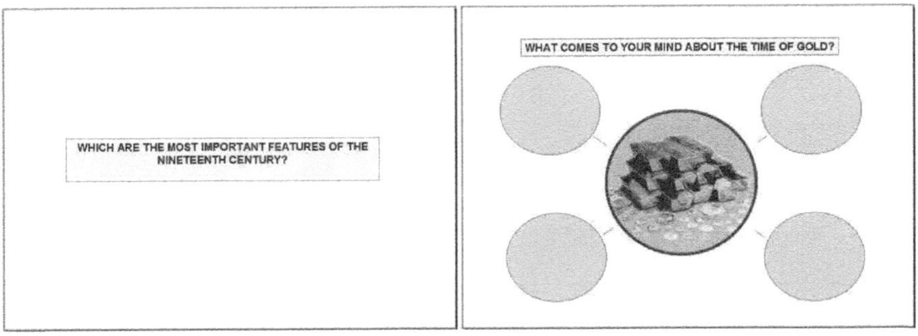

Figure 4.9

The difference in approach was also apparent in the written summary provided to the students at the end of the lesson. For the chronological approach an overview of the six centuries was provided containing the most important years and events. For the imaginative-associative approach the six time periods were presented via their representative image and a number of images with separate blocks of information. The narrative components introduced and used for each time period during the lesson did not occur in these summaries. Refer to Appendix C for the written summaries provided to the students.

The researcher visited the participating classes before the experimental lesson to explain what would be happening and what the aim of that lesson was (not the aim of the experiment, just the information that "a lesson about the history of Aruba" would follow the next week). He also provided some general information about Aruba (where it is situated, what type of island it is) as an introduction to the question that needed to be answered the following week: "How did Aruba—as an island—become what it is today?" The pre-test was also carried out during this lesson. Students were asked to answer a number of general questions about chronology, their knowledge of and affinity with Aruba, and finally to locate twenty-five events from Aruba's history correctly in time. The students were encouraged to complete as much as possible about what they thought they knew. That was not all too easy, because many students, when completing a test about a totally unknown topic that they have not learned about, display the reflex that they consequently could not know anything about it and therefore should not have to fill anything in. In order to encourage the students to write something anyway, there were two example exercises using two events that it was reasonable to assume could be located correctly in time on the basis of general knowledge. This was carried out before the pre-test. In this way it was shown to the students that they perhaps knew more than they thought they did. It was hoped that this would encourage students as much as possible to take the pre-test as seriously as possible. Refer to Appendix D for the questions presented to the students prior to the pre-test and the twenty-five events they had to correctly locate in time during the pre-test.

In the week following the pre-test the lesson was presented by the guest teacher using either approach A or B. The guest teacher used the PowerPoint presentations reproduced in Appendix A and the protocols reproduced in Appendix B for the content of the lesson. A written summary of the subject matter was provided to the students at the end of the lesson (Appendix C). The students were encouraged to read this material before the post-test, which took place a week after the lesson. However, it was apparent that it was not possible to include the grade for this learning as a "real test with a grade," as this would have too great an impact on the existing exam programs in the schools. This had an influence on the extent to which the students seriously studied the summaries. The results section discusses in further detail the extent to which further study of the summary may or may not have played a role.

The post-test was undertaken a week after the experimental lesson. The researcher once again visited all the participating classes to carry out the post-test. The same twenty-five events as used in pre-test were also used on the post-test. This may have caused a test-retest effect, that is, an effect where students score better on the retest due to "learning" during the first test (Cohen et al., 2007; Bryman, 2008). If this was indeed the case, then it may be assumed that it has played an equal role for both groups and therefore has not had an effect on the differences in learning effect measured between both groups.

The post-test was preceded by a number of questions that attempted to establish whether the students had taken an effort to learn more of the material after the experimental lesson: had they read the summary that had been handed out or even attempted to study that material (See Appendix D.5)? The students were encouraged to answer these questions truthfully, as there were no consequences attached had they not done anything with the material.

The post-test was followed by four evaluative questions allowing the students to express their opinion about the lesson and the teacher: did they enjoy the lesson, was the lesson clear, and did they like the teacher (see Appendix D.6)? This allowed any potential differences between the ways in which the students had followed the lessons following approach A or B to be highlighted. Appendix D.4 includes the twenty-five events presented to the students during the pre-test as well as the post-test.

Initial Condition of the Groups

It has been noted before that extra consideration needs to be taken in a quasi-experimental research design for potential differences between the groups. It was therefore investigated how well each group was able to deal with chronology, what their potential prior knowledge was about Aruba and what were their affinity with and attitudes towards being taught on this topic.

Both groups of students were given five multiple choice questions to assess their ability to deal with chronology (see Appendix D.1 for these questions). The first question was aimed at their ability to deal with the general major time periods in history (Antiquity, Middle Ages, Early Modern Time and Modern Time); the second question assessed their skills in determining the length of time periods; the third assessed their skills in locating events within a specific century (e.g., 1672 belongs to the 17th century); the fourth question assessed whether they could determine a sequence; and the last assessed their ability to estimate the length of a time period on the basis of the amount of attention paid to that particular period in a book. The score for each of these five questions was 1 for a correct answer and 0 for an incorrect answer. When totaled, this created a variable for "ability in chronology" expressed as a number between a minimum of 0 and maximum of 5. However, Cronbach's alpha for these five items was 0.324: a low value. The question was therefore whether the same construct was being measured by these five items. The average results for both groups are displayed in Table 4.3.

The average score was high for both groups: students, on average, knew how to correctly answer around four out of the five questions. The low alpha value indicates that nothing more could be established than that both groups were scoring at approximately the same level on five diverse questions about chronology. A t-test did not reveal a significant statistical difference between the averages for the two groups ($t = 0.960$, df $= 259$, $p = 0.338$). The largest difference occurred for the first question, where the average for Group A was 0.86 (SD 0.350) and for Group B 0.73 (SD 0.447).

TABLE 4.3 Average Total Scores to the Five Questions About Chronology

	Group	N	Average	Std. Deviation
Level of mastery of chronology (0–5)	A	134	4.1567	0.96444
	B	127	4.0394	1.01106

Five open-ended questions were presented to the students in both groups to assess their prior knowledge of the history of Aruba (see Appendix D.2). These five open-ended questions were:

- Roughly when was Aruba conquered by the Dutch? The exact answer is 1635, but answers such as the (start of) seventeenth century, or the Golden Century were also considered to be correct. Incorrect answers were those such as "during the time of the East India Company," as this stretches from 1602 to 1799, and this time period was therefore considered too long to be precise enough.
- Who was in power on Aruba when the Dutch conquered the island? The correct answer is the Spaniards. Many students thought that the native islanders were in power. "The English" was also provided as an answer on a number of occasions.
- Which minerals (e.g., coal, oil, or other products found in the ground) have been discovered on Aruba? The correct answers are gold and phosphate. The answers "oil" and "aloe" were deemed as incorrect, as oil has never been discovered on Aruba, although it has been refined there, and aloe is not a mineral.
- What was Aruba used for by the Dutch before 1800? The correct answer is as a "rancho" for horses and goats.
- What was the name of the Aruban leader who achieved a separate autonomous government for the island? The correct answer is Betico Croes.

The score for these five questions was 1 point for a correct answer and 0 points for an incorrect answer. When totaled, this created a variable called "prior knowledge of Aruban history" expressed as a figure between a minimum of 0 and maximum of 5. However, for questions 4 and 5 it was apparent that no students from either group were able to provide the correct answer, and therefore the actual range of this variable was between a minimum of 0 and maximum of 3. The average scores for both groups to these five questions are shown in Table 4.4.

The average score was very low for both groups, and it therefore made little sense to calculate Cronbach's alpha. Prior existing knowledge of

TABLE 4.4 Total Scores on Five Questions on Aruban History

	Group	N	Average	Std. deviation
Knowledge of Aruban history (0–5)	A	135	0.3926	0.56113
	B	131	0.2824	0.50038

Aruban history appeared to have been very low for both groups. A *t*-test revealed there was no statistically significant difference between the groups ($t = 1.688$; df = 264; $p = 0.093$). The largest difference occurred for the first knowledge question (Group A, mean 0.444, SD 0.38; group B, mean 0.388, SD 0.34) about the conquering of Aruba at the start of the seventeenth century. The fact that Group A scored slightly better on this—albeit not statistically significant—could reflect the slightly better ability to deal with chronology in this group that was apparent from the slightly higher scores to the first chronology question. The correct answer to the conquering of Aruba in (the early years of) the seventeenth century rests on a command of Dutch history in general, including, for instance, the history of the East India Company and the West India Company, and therefore does not necessarily rest on any specific knowledge of Aruban history.

A small number of students in both groups (13 in Group A and 8 in Group B) correctly answered "gold" to question 3. Whether or not that indicates any specific knowledge is not clear. They perhaps concluded this from their general knowledge of proverbial gold miners in the New World. For the remaining scores it was apparent that the level of knowledge for both groups was virtually nil. Someone with a reasonable knowledge of Aruban history, for instance schoolchildren from Aruba, would certainly be aware of the name of the Aruban leader, Betico Croes. This did not appear to be the case for any of these Dutch students in either group. It may therefore be concluded that there was no specific prior knowledge of Aruban history present in either group.

Students in both groups were asked whether they had ever visited Aruba or one of the other islands of the Netherlands Antilles, in order to assess their potential affinity with the topic. In both groups 13 students answered "yes" to this question with the remainder answering "no." In addition, they were asked whether they knew anybody from Aruba or one of the other Antilles, and if so whether they "simply knew them" or "knew them very well." In Group A 35 students answered that they "simply knew" people from the islands; for Group B there were 40 students. For Group A 18 students replied that they knew people from the islands "very well" and for Group B there were 16 students in this category. These differences are very small and did not appear to be statistically significant.

The students in both groups were asked whether Aruban history was an important topic to the Dutch, in order to assess their attitudes toward the subject. Students were able to answer "yes" or "no" to this question, although some wrote "a little," in addition. For the scoring to this question, 0 points were allocated for "no," 1 for "a little" and 2 points for "yes." In addition, the students were asked whether they would enjoy having a lesson

TABLE 4.5 Attitude Towards (Lessons About) the Topic "History of Aruba"

	Group	N	Average	Std. deviation
Attitude towards topic Aruba (0–4)	A	124	2.2419	1.38147
	B	116	2.2586	1.42719

about Aruban history. Answers to this question ranged from "no, wouldn't enjoy" (0 points), "I don't know" or "doesn't really matter to me" (1 point), and "yes, would enjoy" (2 points). Summing the points to these two questions created a variable "attitude toward the topic of Aruba" which varied in scores from 0 to 4. The average scores are presented in Table 4.5.

For both groups the average score was higher than 2, implying, in general, a positive attitude toward the topic and lesson experiment. The differences between the groups were very small and were not statistically significant according to a *t*-test ($t = 0.092$; df = 238; $p = 0.927$).

Conclusion. The differences between the two groups appear to have been very limited up to the point of the pre-test. For the group composition, it was apparent that the number of advanced level students in Group A was slightly higher than in Group B, although this difference was not statistically significant. Group A appeared to have a slightly enhanced baseline status compared to Group B on a number of other characteristics, but these differences were not statistically significant either. The general ability to deal with chronology appeared to be slightly better in Group A compared to Group B, which concurs with the fact that slightly more students in Group A were able to locate the conquering of Aruba correctly in time. The differences between the groups were small, but insofar as they could have an influence on the results of the pre-test and post-test, it was anticipated that Group A would perform better than Group B.

The Test Used During the Pre-Test and Post-Test

During the pre-test and post-test, students had to indicate when 25 events of out Aruban history had taken place (see Appendix D.4). Nine of these twenty-five items were explicitly referred to during the lesson by the teacher, in approach A as well as in approach B; the remaining sixteen had to be linked to the correct time period or century by the students through logical reasoning using their historical reference framework. This difference did not have any effect at the pre-test, as students did not, at that stage, have any reference framework for the history of Aruba at all. So during the pre-test, all location in time had to take place based on logical reasoning using a general historical reference framework. However, in the analysis of the post-test a distinction was drawn between the category of reproduction (events referred to during the lessons) and the category of application (events not referred to during the lessons).

Another distinction important to this experiment is the presence of a keyword in the description of an event that also occurred in the name for a time period in the imaginative-associative approach. This applied to eight events: "Indians" for events 15 and 17, "horses" for events 3, 17 and 18, "pirates" for event 8, and "oil" for events 6, 16 and 22. This distinction had no effect on the pre-test, however for the post-test students in Group B would be able to answer on the basis of word recognition. However, there were two instances where this would lead students down the wrong track: event 15, which refers to an "Indian" chief, could erroneously evoke the "time of the Indians" and event 17, which refers to both Indians and horses, while the time period referred to is neither the Time of the Indians, nor that of horses and pirates. Word recognition could lead to the correct location of a time period in six other instances for Group B. We could investigate to what extent keyword recognition played a role for students in Group B by making a distinction in the analysis of the post-test results between events that did and those that did not include a keyword in the description. An overview of the test items by the various categories is shown in Table 4.6.

Scoring the Tests

The test form indicated that students could use various options for locating an event in time, including: numbered years, centuries or descriptions such as "the time of..." Students were also able to tick the box "I don't know" where applicable. The latter was included to prevent students from entering all kinds of wild guesses with years and centuries simply to write something down, which would have had a highly negative impact on the reliability of the data. However, one consequence of this was possibly that students were making use of this option too readily, even in the case of events the timing of which they could reasonably be expected to work out.

TABLE 4.6 Survey of Test Items of Different Categories in the Pre-Test and Post-Test

	Reproduction	Application
with keyword	6, 8	3, 15, 16, 17, 18, 22
without keyword	2, 5, 9, 12, 14, 24, 25	1, 4, 7, 10, 11, 13, 19, 20, 21, 23

For instance, it is obvious that the construction of an airport and building a shopping mall will have taken place in the twentieth century (events 23 and 13). Nevertheless more than half of the students indicated at the pre-test not to know this or entered a wrong answer for the question.

The same rules were applied for scoring both the pre-test and the post-test. A completely correct answer was scored as 2 points, a completely incorrect answer or "I don't know" as 0 points. There was a category in between these two of "somewhat correct" which was awarded 1 point, where the answer was heading in the right direction but was not completely correct. This category was primarily used for the eleven events of the twentieth century (numbers 2, 5, 6, 7, 10, 13, 14, 16 and 22). An additional requirement was imposed on these questions; simply reporting the "twentieth century" only resulted in 1 point. This was put in place as the twentieth century consisted of two different eras in the imaginative-associative approach (Time of Oil and Time of Tourism), from which the correct one had to be selected. In contrast to this, the seventeenth and eighteenth centuries were considered as a single period within the imaginative-associative approach (Time of Horses and Pirates), and therefore events from this period (numbers 3, 8, 18 and 19) could be scored as 2 points for both the "seventeenth century" and the "eighteenth century" (or years from that period). A description such as the "time of the East India Company" was also scored this way in this case, as this coincided almost in full with this period. Descriptions of time periods referring to a period of more than a hundred years that included the correct time point were scored as 1 point, with the exception of the periods 1600–1800 and the one prior to 1500.

Events creating a "modern impression," such as the construction of a shopping mall (13), building an airport (23) or creating an installation to distil water from sea water (24) were scored as zero if they contained "the present" or "now" or a year after 2000, given that this refers to associations with the present day rather than the historical past. Answers such as these do not indicate any historical awareness, nor any attempt to locate events in a historical time period. The reverse applies to event 4 (the disappearance of the American girl, Natalee Holloway in 2005). Because this event could only have been known from the news, students had to answer with "now" or in time after 2000. All other dates were scored as zero for this question.

Event 1 had to be located in the sixteenth century or the Time of the Useless Island, which refers to the time following the year in which the Spaniards deported the entire population of Aruba. Event 9 refers to the gold fever of the nineteenth century and therefore should be designated as belonging to the nineteenth century or the Time of Gold. Event 11 refers to the period where there was no influence of Christianity on Aruba yet; therefore any answer indicating the period before the sixteenth century (or the Time of the Indians) would be scored as 2 points. Answers including reference to the sixteenth century received a score of 1, as it could not be ruled out that some Indians had not yet converted to Christianity during that century. Event 12 had to be designated to have occurred in the sixteenth century or the Time of the Useless Island; this refers to the deportation of the Indian population by the Spaniards during that period. Event 12 had to be designated to have occurred in the sixteenth century or the Time of the Useless Island, as Christianity had apparently already arrived on the island (therefore after the arrival of the Spaniards), but not all Indians had been converted yet (therefore before the Dutch conquest). Event 20 refers to mining and therefore gold prospecting. The correct answers would include the nineteenth century or the Time of Gold. The same applies to event 21. Event 25 refers to the start of the colonial period, the time after the West Indies Company, and therefore should be located at the start of the nineteenth century. Around 1800, nineteenth century, or the Time of Gold received a 2-point score. An overview of the scoring methods applied can be found in Appendix E.

Twenty combinations of pre-tests and post-tests, chosen via a random number generator, were scored by a second, independent rater. The observed inter-rater agreement in the scores amounted to 98.2% for the pre-test and 93.8% for the post-test with a Cohen's kappa of 0.98 and 0.96 respectively.

Test Reliability

The internal reliability of the test (the extent to which the items were measuring the same construct, i.e., in this instance, the ability of the students to place events in time correctly) was determined for the pre-test and the post-test. In the first instance Cronbach's alpha was 0.807 and in the latter 0.880, both of which are high values. A limit of 0.80 is usually employed in social science research to indicate that a test is demonstrating high internal reliability, although some researchers also accept lower values (Cohen et al., 2007; Bryman, 2008). We investigated whether certain items had an undue influence on the internal reliability of the test by iteratively removing a single test item and re-assessing Cronbach's alpha for the pre-test and post-test. Cronbach's alpha was shown to be consistently lower for both the pre-test and post-test when items were removed with five exceptions: the alpha value was higher when items 1, 4, 12, 15 and 19 were removed from the pre-test, and when items 1, 4, 17, 19 and 25 were removed from the post-test. Therefore items 1, 4 and 19 had a negative influence on Cronbach's alpha for both the pre-test and post-test. An explanation for this for item 1 could be that an "empty island" invokes strong associations with the period prior to the European colonization, which, of course, is incorrect. Item 4 is different in nature to the others, as this involves knowledge of current affairs. Item 19 is a tricky item requiring both thorough knowledge and making numerous associations: the Dutch govern the island, they are Protestants, and therefore do not want to admit any Catholic priests, yet the Indian population of Aruba had been converted to Catholicism previously (by the Spaniards) and therefore had to worship in secret. This situation only applies to the 17th and 18th centuries, but that requires quite a degree of reasoning to work out.

During the pre-test, items 12 and 15 invoked associations with the slavery period and age of discovery respectively, both of which are incorrect. Item 17 was one of the most difficult at the post-test, as it fell during the interim period between the end of Spanish rule and the point at which the Dutch took over. Item 25 was often associated with the time at which the Dutch first conquered Aruba (and therefore made up the balance from what was already present), although in reality this occurred at the start of the nineteenth century—as was reported during the lessons. The items that had a negative effect on Cronbach's alpha were items that invoked incorrect associations during the pre-test, and those that required more than simple associations with characteristics of particular ages and periods during the post-test.

When all five items with a negative effect on Cronbach's alpha at the pre-test were removed, Cronbach's alpha for the remaining twenty items was 0.827. For the post-test Cronbach's alpha was 0.894 when the five items with a negative effect were removed. Although these are higher values than for the entire 25 items in the test, the Cronbach's alpha values were not higher to such a degree that it would justify excluding these items from the analyses of test results.

The internal test of reliability of the four previously described subcategories of the items (see Table 4.6) was determined for the post-test. It would not make sense to assess this for the pre-test, given that the categorization relates to the subject matter presented to the students during the intervention. The alpha values for the various subcategories may be found in Table 4.7. The internal reliability of all four subcategories appeared to be at an acceptable level.

TABLE 4.7 Cronbach's Alpha for Subcategories of Test Items on the Post-Test

Subcategory	Cronbach's alpha
with keywords	0.782/0.817[a]
without keywords	0.806
reproduction	0.758
application	0.813

[a] The value of alpha was calculated for all eight items with keywords, and for the six items which led students in Group B towards correct answers based on recognition of keywords. The second value applies to this last case.

Results

The Pre-Test

A difference in scores between Group A and Group B at the pre-test would mean that one group was at a greater advantage to locate events in time correctly than the other group. This would need to be taken into consideration in interpreting the results of the post-test. As previously explained, the maximum score was 2 for each item. The high Cronbach's alpha value justified totaling the scores to all the questions from the pre-test, as the same construct was being measured to a significant extent by these questions. Therefore the score for the pre-test was determined by totaling the scores to each separate question, creating a potential maximum score of 50 (25 times 2). It was apparent from the results of the pre-test that the average score for all 267 students was 8.60, the maximum score 29 and the minimum score 0. Table 4.8 shows the average results split across Groups A and B.

It is apparent when comparing the data from the two groups that there is a considerable difference, which was shown to be statistically significant when a *t*-test was applied ($t = 3.550$; $df = 265$; $p < 0.001$). The expectation that Group A would score higher than Group B was confirmed by this result. The difference was, however, considerably greater than expected on the basis of the small differences between the groups previously noted. Figure 4.10

TABLE 4.8 Average Scores on the Pre-Test

	Group	N	Average	Std. deviation
Pre-test (0–50)	A	135	10.0000	6.94542
	B	132	7.1667	6.05593

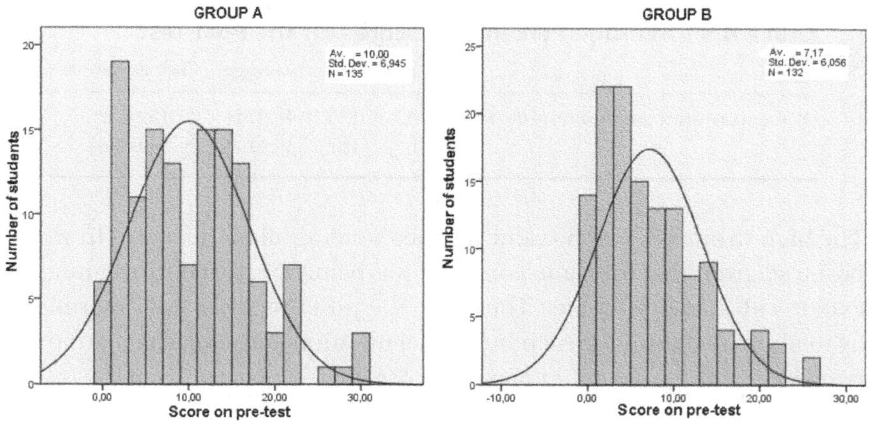

Figure 4.10

shows how the scores are distributed for both groups: in both instances there is a "skewed" distribution to the left of the histogram, which is referred to as a "positive skew." For Group A this positive skew amounts to 0.640, and for Group B 0.997, meaning that the number of scores in the left area of the histogram are relatively greater in Group B than in Group A.

Sixty-one students scored zero, one or two points, probably indicating that they had indeed often simply yielded to the temptation to answer "I don't know," even if that did not tally with what was actually the case. The number of students scoring zero to two points at the pre-test amounted to 25 in Group A (18.5%) and 36 in Group B (27.3%). This is a considerable difference. An explanation for these differences between the two groups could be that Group A was better able to locate events correctly in time during the pre-test than Group B. Group A, therefore, had a considerably better starting point than Group B, which is why the pre-test score was used as a potential influential co-variable when interpreting the results of the post-test. It is assumed that the limited differences between Groups A and B previously demonstrated could be considered as a cause for the difference at the pre-test, and therefore that these limited differences are taken into account when the pre-test result is included into the analyses of the post-test.

The Post-Test

The total scores were transformed into percentage scores for calculating the post-test results (the percentage out of a maximum score of 50). This was undertaken in order to facilitate the comparison between scores in subcategories of questions, which were also expressed as percentages.

TABLE 4.9 Average Percentage Scores on the Post-Test

	Group	N	Average	Std. deviation
Percentage score on post-test (0–100)	A	135	40.0741	18.37326
	B	132	49.5758	24.06864

The high Cronbach's alpha value justified totaling all of the scores from the post-test, given that the same construct was being measured to a significant extent with these questions. Therefore the post-test score was determined by totaling the scores for each individual question and then calculating the percentage of the maximum possible score. It was apparent from the results of the post-test that the average scores for all 267 students was 44.8%, the maximum was 90.0% and the minimum score 0%. Table 4.9 shows the average results for both Group A and Group B.

The difference between the two groups is now reversed compared to the difference that turned up at the pre-test: where Group A scored higher at the pre-test, Group B now has the higher score. This difference is considerable and statistically significant according to a *t*-test ($t = 3.620$, $df = 245.041$, $p < 0.001$). Figure 4.11 shows how the scores were distributed across the two groups: for Group A the scores are approximately normally distributed with a single outlier on the left; there is a skewed distribution of the scores in Group B to the right. The skew has changed in respect to the pre-test in Group A from 0.640 to 0.232 and from plus 0.997 to minus 0.452 in Group B. This implies a much greater shift in Group B than in Group A. At the pre-test there was a skew in both groups towards lower scores, which

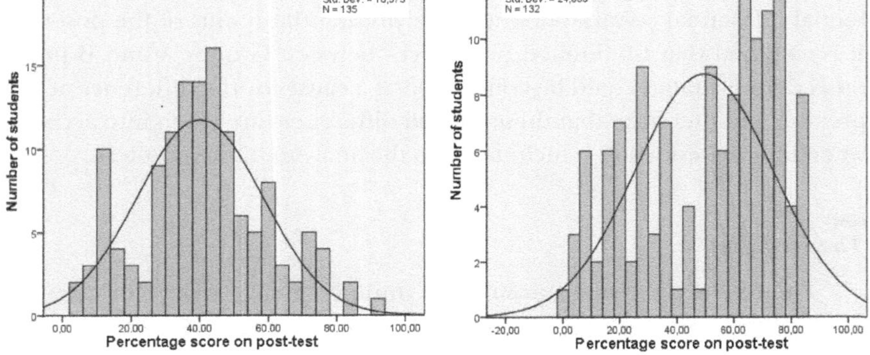

Figure 4.11

was more pronounced in Group B than in Group A. For the post-test there was still a slight skew for Group A towards lower scores, but less than at the pre-test. For Group B there was a reversal from lower scores to higher scores (that is from positive skew to negative skew). However, it is noteworthy in Group B that the histogram shows two peaks instead of one, as is to be expected normally. This could mean that approach B apparently either worked well (for most students), or did not catch on (for some students), the contrast between lower and higher results being more outspoken than in a normal pattern. The explanation of this phenomenon is as yet unclear.

When comparing scores for the 25 questions separately (Appendix F) it may be seen that on the pre-test Group A scored higher than Group B in 24 out of 25 instances. This was reversed by the results of the post-test, where Group B scored higher than Group A in 23 out of 25 instances. A paired *t*-test was used to determine which items had significantly higher scores for the post-test compared to the pre-test. In Group A this amounted to sixteen items and in Group B to twenty-two items. In Group B the t-value for the difference was higher in twenty-three cases in comparison to those of Group A (Appendix G).

Prior to the twenty-five test items requiring students to locate events in time correctly, the students were asked at the post-test whether they had read or studied the written summary provided the week before during the time between the experimental lesson and the post-test (see Appendix D.5). The number of students stating that they had studied the summary was very small for both groups. Furthermore, there is the *post hoc* question of whether a clear distinction can be drawn between the reading and studying categories. A nominal variable was therefore created on the basis of these questions: studied/learned, "Yes" or "No." This appears to have had a considerable effect on the post-test scores. The number of students in both counts who stated that they read or studied the summary can be seen in Table 4.10, including the average scores for the subgroups distinguished in this way.

TABLE 4.10 Average Results on the Post-Test in Subcategories of Students Who Said to Have Read/Studied or Not to Have Read/Studied the Summaries

	Group	N	Average	Std. deviation
Post-test for subgroup NOT read/ studied (0–100)	A	40	30.90	15.60
	B	51	34.78	22.09
Post-test for subgroup read/studied (0–100)	A	95	43.94	18.15
	B	80	59.30	20.19

The significant effect of cognizance of the written summary is apparent from the data in Table 4.10, with the greatest effect demonstrated in Group B. It is apparent in both instances that the average result is higher for Group B than Group A; however, this difference was considerably greater in the case of the group of students who claimed to have read or studied the summary in comparison to those who claimed not to have done so. A t-test revealed that the difference between Groups A and B for those students who stated *not* to have studied the summary was not statistically significant ($t = 0.982$, df = 88.129; $p = 0.329$), while the difference between Groups A and B was statistically significant for those students who *did* study the summary ($t = 5.298$, df = 173, $p < 0.001$). It therefore seemed sensible to draw a distinction between those students who said they had and those who said they had not read or studied the summary. This created four conditions after the interventions:

- Students who had followed the lesson according to the chronological approach and had read or studied the summary (Condition 1);
- Students who had followed the lesson according to the imaginative-associative approach and had read or studied the summary (Condition 2);
- Students who had followed the lesson according to the chronological approach and had not read or studied the summary (Condition 3);
- Students who had followed the lesson according to the imaginative-associative approach and had not read or studied the summary (Condition 4).

The total scores and scores to the question subcategories are summarized in Table 4.11 for all students and for students belonging to the different conditions.

Statistically significant differences on the basis of the t-tests between the results of Group B compared to those of Group A have been highlighted in the table with an asterisk. For each of these significant differences, Group B always scored higher than Group A. No significant differences were demonstrated for the subcategory “application without keywords” for the assessment across all students; however this difference *was* significant for those students who claimed to have read or studied the material. The difference between Group A and Group B was not statistically significant in any instance where students said they had *not* read or studied the material.

TABLE 4.11 Survey of Percentage Scores on Post-Test in Diverse Subgroups

Group		All post-test items			Post-test items without keywords		
		Total	Repr.	Appl.	Total	Repr.	Appl.
All students							
A	N	135	135	135	135	135	135
	Average	40.07	37.49	41.53	40.19	36.24	42.96
	Std. deviation	18.37	21.51	19.28	18.28	21.98	19.63
B	N	132	132	132	132	132	132
	Average	49.58*	48.95*	49.93*	47.93*	45.29*	49.77
	Std. deviation	24.07	27.30	23.86	22.72	25.79	23.74
Students who said to have read/studied the summaries							
A	N	95	95	95	95	95	95
	Average	43.94	41.99	45.03	43.44	40.83	45.26
	Std. deviation	18.15	22.03	18.76	18.43	22.68	19.40
B	N	80	80	80	80	80	80
	Average	59.30*	59.10*	59.41*	56.54*	54.46*	58.00*
	Std. deviation	20.19	23.94	20.22	18.89	23.47	20.29
Students who said NOT to have read/studied the summaries							
A	N	40	40	40	40	40	40
	Average	30.90	26.81	33.20	32.50	25.36	37.50
	Std. deviation	15.60	15.96	18.12	15.59	15.76	19.32
B	N	52	52	52	52	52	52
	Average	34.62	33.33	35.34	34.67	31.18	37.12
	Std. deviation	21.90	24.85	21.70	21.85	22.81	23.25

* Statistically significant difference between group B and A ($p < 0.005$), see also Appendix H.

The effect of the four conditions on the results of the post-test can only be truly determined if corrected for the effect of the difference on the pre-test, given that there were significant differences apparent between the groups at the pre-test. To what extent can the difference in the results of the post-test be explained by the four different conditions (that is the intervention) and to what extent is this due to the differences between groups existing prior to the intervention, as apparent from the pre-test? An analysis of covariance was used to determine what the effect of the intervention had been using teaching method applied and the read/studied factor as independent variables (as these determined the condition), the result of the pre-test as a co-variable, and the results of the post-test as the dependent variable. This ANCOVA demonstrated significant effects on the results of the post-test for both the teaching method applied ($F(1,261) = 28.969$, $p < 0.001$, $\eta_p^2 = 0.100$), as well as for the read/

studied factor ($F(1,261) = 54.708$, $p < 0.001$, $\eta_p^2 = 0.173$), and the pre-test ($F(1,261) = 56.308$, $p < 0.001$, $\eta_p^2 = 0.177$). There was also a significant interaction effect between the teaching method applied and the read/studied factor ($F(1,261) = 10.832$, $p = 0{,}001$, $\eta_p^2 = 0.040$); that is, whether or not the summary was studied or read had a significant influence on the effect of the teaching method applied. This effect was greater for Group B than Group A: the result corrected for the effect of the pre-test on the results of the post-test was 31.73% (SD 2.77) for Group A for the group who did not read or study the summary and 41.13% (SD 1.83) for the group who had. For Group B these figures were 36.61% (SD 2.46) and 61.06% (SD 1.97), respectively. The two different teaching methods, as well as the differences in whether or not the summary had been read or studied (the interventions) therefore had a demonstrable effect on the results of the post-test, even when the differences at the pre-test were taken into account. These differences were caused by differences *between* groups and not by potential deviations in the group averages caused by the differences *within* groups.

The effect of the pre-test and that of the interventions on the results of the post-test could also be demonstrated in another way, that is, by determining the Pearson correlations between pre-test and post-test results. If these correlations were both considerable and significant, then it could be said that there was a relatively strong relationship between pre-test and post-test results. If they were smaller and not significant, then it could be assumed that the post-test score was due to something other than the pre-test, that is, it could only be attributable to the intervention. The Pearson correlations have been summarized in Table 4.12 for all students and separately for the four conditions, totals, as well as being split across the six different subcategories of test items.

For Group A the correlations for all students and those students who said they had read or studied the summary are greater for all subcategories of test items than those for Group B, as well as being statistically significant in all cases for Group A ($p < 0.001$, highlighted in the table with "*" and bold figures); for Group B this is only the case for both of the subcategories "application" and the subcategory "total, items without keywords, reproduction" for all students taken together. However, the correlation for the latter was significantly weaker than that for Group A. In other words, for all students taken together and for those students who said they had read or studied the summary there was a stronger correlation between the pre-test and the post-test in Group A compared to Group B, in other words, it may be assumed that for all students in Group B the result of the post-test is to a greater extent attributable to something other than the pre-test, that is the intervention. This is most apparent for the group of students who said they

TABLE 4.12 Pearson Correlations Between Pre-Test and Post-Test Results

	All post-test items			Post-test items without keywords		
Group	Total	Repr.	Appl.	Total	Repr.	Appl.
All students						
A Pearson correlation	**0.543***	**0.385***	**0.566***	**0.501***	**0.330***	**0.535***
p	< 0.001	< 0.001	< 0.001	< 0.001	< 0.001	< 0.001
N	135	135	135	135	135	135
B Pearson correlation	0.298	0.255	**0.305***	**0.311***	0.242	**0.321***
p	0.001	0.003	< 0.001	< 0.001	0.005	< 0.001
N	132	132	132	132	132	132
Students who said they had read/studied the summaries						
A Pearson correlation	**0.545***	**0.392***	**0.565***	**0.516***	**0.361***	**0.539***
p	< 0.001	< 0.001	< 0.001	< 0.001	< 0.001	< 0.001
N	95	95	95	95	95	95
B Pearson correlation	0.164	0.138	0.164	0.146	0.124	0.131
p	0.146	0.222	0.146	0.197	0.275	0.248
N	80	80	80	80	80	80
Students who said they had NOT read/studied the summaries						
A Pearson correlation	0.435	0.199	**0.487***	0.350	0.029	**0.463***
p	0.005	0.218	0.001	0.027	0.860	0.003
N	40	40	40	40	40	40
B Pearson correlation	**0.598***	**0.506***	**0.618***	**0.630***	**0.500***	**0.664***
p	< 0.001	< 0.001	< 0.001	< 0.001	< 0.001	< 0.001
N	52	52	52	52	52	52

* Correlation statistically significant.

had studied or read the written summary. The opposite is true for those students who stated not to have read or studied the summary: the correlation for this group between pre-test and post-test is stronger in Group B compared to A, and presumably therefore the effect of the intervention is proportionally less. These results reflect those from the ANCOVA: the effect of the imaginative-associative approach is most clearly apparent for those students who said they had read or studied the written summary.

Evaluation of Both Approaches by the Students

One of the questions that had to be answered by students prior to the post-test was the question of whether or not the students had paid attention during the lesson (answer "yes" = 1 or "no" = 0). The average score to this question for Group A was 0.85 and 0.92 for Group B; this is a small differ-

TABLE 4.13 Appreciation by Students of the Lesson about Aruba and the Guest Teacher

	Group	N	Average	Std. deviation
Enjoyed lesson (0–5)	A	134	2.18	0.93
	B	130	2.37	0.92
Found lesson comprehensible (0–5)	A	134	2.51	0.96
	B	130	3.00	0.77
Liked teacher (0–5)	A	134	2.75	0.69
	B	130	3.03	0.61

ence that was not statistically significant ($t = 1.639$, df 252.523, $p = 0.102$). This indicates that probably any potential difference in evaluation and/or difference in performance cannot be attributed to the extent to which students were paying attention during the lesson.

The students were able to provide their opinion about the lesson they had attended at the end of the post-test (see Appendix D.6). They were asked whether they had enjoyed the lesson and whether they found it comprehensible. They were also asked to provide their opinion about the teacher: did they like the teacher? These three questions were scored on a five-point scale ranging from 0 (not enjoyable at all, not clear at all and not very nice at all) to 4 (very enjoyable, very clear and very nice). The averages for this assessment are shown in Table 4.13.

The difference between Groups A and B for the question "enjoyed the lesson" was not statistically significant ($t = 1.678$, df = 262, $p = 0.95$). Students in Group A scored significantly lower than students in Group B in terms of finding the lesson comprehensible ($t = 4.623$; df = 253.229, $p < 0.001$). It is apparent that this was also transferred to the evaluation of the teacher; although the difference here was somewhat smaller it still was statistically significant ($t = 3.461$, df = 262; $p = 0.001$). Both groups, however, rated that they quite liked the teacher.

Discussion of Results

It was apparent from the discussion of the differences between Groups A and B in Section 1 that Group A had an edge over Group B (a slightly greater number of advanced students, and a slightly better knowledge of chronology), and it was therefore to be anticipated that Group A would outperform Group B on the pre-test and post-test. This was confirmed for the pre-test, but the difference was considerably greater than had been anticipated on the basis of the very limited differences demonstrated at baseline: Group A scored considerably higher than Group B. This difference was also statistically significant. It is therefore apparent that Group A was at a higher level than Group B prior to the intervention. This was reversed after the intervention: Group B scored considerably higher than Group A, and this difference was also statistically significant. The expected results were therefore not confirmed here. This implies that Group B had formed a better reference framework during the intervention than Group A, and this enabled better results to be realized. This result cannot easily be attributed to anything other than the intervention, given the factors that were taken into account during the analysis of the data. The results from the analysis of variance clearly demonstrated that the effect of the pre-test results (i.e., a higher initial level) was considerable, but the difference between the groups could not be solely attributed to the pre-test levels. According to this analysis, the differences must have been caused by the intervention, at least partly, and this is more evident because the results of the post-test was reversed compared to those of the pre-test.

The distribution of the results as shown in the histograms in Figures 4.10 and 4.11 points in the same direction. At the pre-test there is a positive skew for both groups as expected. In other words, there are relatively speaking a lot more outcomes in the area of the lower scores: the students knew very little or nothing at all about the subject. For Group B this skew was more pronounced than for Group A, which concurs with the lower average result for Group B at the pre-test. In the results of the post-test there is still a slight positive skew for Group A, possibly caused by a single outlier around the 10% area. The remaining distribution of the results for Group A was reasonably "normal." However, for Group B the image is clearly different. The positively-skewed distribution was transformed into a negative skew, suggesting a clear reversal. The skewed distribution in the region between 60% and 80% demonstrated that a considerable proportion of the students in Group B achieved a relatively high score. However, in contrast to this, the other side of the histogram also demonstrates a number of outliers. It is noticeable that in the mid-region—where the highest number of scores can be expected for a normal distribution—there is a drop in the number of scores for Group B. Apparently the imaginative-associative approach either worked very well or it did not work; there are relatively few average scores.

There were two "media" within the intervention: the lesson taught either with the imaginative-associative or chronological approach, and the written summary provided in accordance with these two approaches. It was apparent from the analyses that whether or not the summary was studied had a relatively large effect on the post-test results. Apart from the difference between the numbered centuries and time periods with imaginative names, the written summaries also displayed the difference between, on the one hand, a systematic summary of chronological information, and on the other more image material and separate chunks of information with a less systematic association (see Appendix C). One could wonder what has been the effect of the image material in the summary—which was compatible with an orientation in time according to the imaginative-associative method. It could be assumed that the more attractive looking imaginative-associative summary could have had a motivating effect on students to take note of the material at all. This, however, was not confirmed by the numbers of students in Groups A and B who said they had read or studied the summary: this number was 95 students (70.4%) in Group A and 80 students (61.1%) in Group B. This suggests a greater level of industriousness and motivation in Group A, which is compatible with their better baseline situation. Yet despite this, Group A scored less well at the post-test. The extent of the influence of the image material cannot be determined with these results.

The difference between both groups in cases where the summary had been read or studied was considerably greater than the difference between the groups where it had not been read or studied, and greater than the difference in general between both groups. The difference between the groups of students who said they had not read or studied the material was not statistically significant. This may mean that the lesson did not have enough impact by itself to generate a difference between the two approaches. However, the group of students who had not read or studied the material was small, and the smaller the group, the harder it is to demonstrate a significant difference, as greater "demands" in terms of levels of significance are imposed on the results (Cohen et al., 2007). It may also be assumed that those students who said they had not read or studied the material were less industrious or less interested, and so the lesson could also have had less of an impact on these students. Either way, no difference between the approaches was demonstrated for the minority of students who said they had not read or studied the written summaries. It may therefore be concluded that the impact of lessons on learning was somewhat limited. This is understandable given that they only involved one presentation, which for both approaches involved a long lecture, which is not an optimal environment to promote learning. It seems therefore justified to assume that within a normal teaching environment, where space has been provided for a variety of methods, students attend to their homework and are able to thoroughly prepare for tests, an imaginative-associative approach would lead to a greater degree of learning than a chronology-based approach.

What was the extent of the effect of the presence of certain associative words in the names for time periods and the description of events? It appears logical that students in Group B were relatively easily able to associate descriptions of events in which, for instance, the word "oil" appeared, with the "Time of Oil." Group A were not able to use these straightforward associations. To what extent was the greater learning effect in Group B attributable to this type of mechanism? If there was a large influence, then there would have been less evidence of real historical orientation and a greater degree of identification on the basis of word recognition. As has been previously noted, word recognition was not always positive for students in Group B. In the case of two events words were used in the descriptions that led them down the wrong path, namely for event 15 that refers to an "Indian" chief, incorrectly evoking the Time of the Indians, and event 17 that refers to both Indians and horses, though the event belongs neither to the Time of the Indians, nor to that of Horses and Pirates. Event 17 is also one of the events where Group A scored higher than Group B (see Appendices F and G). However, in six other cases associations like this did benefit students in Group B. It would not be entirely justified to ignore these cases completely, given that the imaginative-associative method of orientation in time is partly based on this type of association and therefore provided a benefit to Group B with good reason. It was, however, also noted that when the eight events with keywords were removed, Group B still consistently scored higher than Group A; this difference was also statistically significant. Nevertheless, the difference was somewhat smaller than the one for the post-test results across all 25 events: Group A achieved results that were virtually identical to the results across all 25 events, whereas Group B achieved scores just below the total result for the 25 events, which was of course to be anticipated. A paired *t*-test of the differences between pre-test and post-test results shows that for the eight cases in which Group B recorded significant progress and Group A did not, there was only one in which this progress could be attributed to the recognition of a keyword (item 8), while there were four cases of items from the category "application without keywords" (items 4, 7, 10 and 11). The greater learning effect for students in Group B was therefore only to a limited extent based on the recognition of keywords from the names of the time periods. Real "associative" factors have probably played a greater role.

If students remembered events referred to in the lessons and were able to locate these correctly in time, then this could be explained as the ability to recall knowledge, or as knowledge reproduction. For those events that did not occur in the lessons, the process of locating them correctly in time involved the application of knowledge, which is a higher category of learning objectives than reproduction. It was observed that students in Group B scored higher results on the post-test compared to students in Group A in all instances, although the difference for the "application" level was only significant when assessed for those students who said they had read or studied the material (see Table 4.11). This means that the imaginative-associative approach was not only better at helping students to remember events presented in the lesson, but also in locating "new" events. The question is to what extent the latter was determined through the occurrence of a keyword in the name for a time period and in the description of an event. It is apparent from the data in Table 4.11 that this was the case to a limited extent for the total group and the group who said they had read and studied the material. When the test items with keywords are removed, Group A scores approximately the same as for all 25 test items. This is to be expected, as the keywords for Group A, were not, of course, really keywords at all. For Group B we can see that when the test items with keywords are removed there is a slight reduction in scores compared to those for all 25 events, particularly in the reproduction category. The difference between the two groups is therefore reduced, which indicates that word recognition indeed played a role for Group B, but not nearly to such an extent that it explains all of the difference. True associative factors probably played a role for Group B in both the reproduction as well as application of knowledge. Extending this study by (for instance) using thinking aloud methods or interviews with students would clarify whether these associations are occurring and, if so, to what extent.

It is notable that the reproduction scores are lower in all instances compared with application for both Group A and Group B, while reproduction of the subject matter is considered to represent a lower level. The reason for this is difficult to ascertain. Perhaps being confronted with subject matter in a single lesson in association with a less thorough cognizance of the written summary of that information was too superficial to generate proper reproduction of the material.

The results of both the ANCOVA and the comparison of the Pearson correlations make it more than likely that the differences between Group A and Group B are actually attributable to the intervention and not to other factors. The ANCOVA demonstrated a significant effect for both the teaching method applied, as well as whether or not the material was read or studied once the effect of the pre-test had been corrected. In other words, significant differences arose between the groups as a result of the intervention that cannot be explained by odd distributions of differences within group or by a better initial situation.

This was confirmed again by the Pearson correlations between the pre-test and the post-test results. It was apparent from Table 4.12 that the correlations between the pre-test and post-test results for all students, as well as just those who said they had read and studied the summary, was greater in Group A for all subcategories of test items than in Group B. It may be assumed from this that the result in Group B for those cases was determined to a greater extent by something other than the pre-test, that is, by the intervention (so the imaginative-associative approach in Group B). This applies to both reproduction and application, and both for those situations where keyword recognition could be used, as well as those where this was not possible. For the small group of students who said they had not read or studied the material, the correlation between pre-test and post-test results was greater in Group B than Group A. This underlines once again that the effect of the imaginative-associative intervention could not be demonstrated for the group of students who said they did not read or study the summary. It was noted previously that this is not so remarkable in the case of attending to one lesson in the form of a long lecture supported by PowerPoint, particularly if it is assumed that those students who had not taken the trouble to review the summary were probably also the less able and less motivated students. However, the latter cannot be established on the basis of the data assessed during this experiment. The group who said they had not studied or read the summary was noticeably small, and it is therefore difficult to demonstrate significant differences between the subgroups.

In terms of the students' evaluation of both approaches, it was apparent that the students in both groups enjoyed the lesson and quite liked the teacher. The students in both groups stated that they had paid proper attention during the lesson. The ratings for both groups were approximately the same in this respect. However, students in Group B believed to a much greater extent that the lesson had been comprehensible. This agrees with the fact that they also achieved a much greater learning effect. The standard deviation for the responses to the question regarding clarity was much lower for Group B than Group A, which indicates that there was a greater degree of agreement between the students regarding the clarity of the imaginative-associative approach. This cannot be attributed to a difference in the style of the teachers' presentations, given the same guest teacher presented the lesson to the classes in the parallel groups.

Conclusions

A parallel lesson was presented to two comparable groups of students in order to determine whether orientation in historical time is supported to a greater extent by chronological teaching methods or to a greater extent by imaginative-associative teaching methods. Comparability of the two groups was essential in order to be able to draw conclusions. It was demonstrated that there was a significant degree of comparability between the two groups, although the group confronted with the chronological approach demonstrated higher results on the pre-test, concurring with a number of different characteristics of that group (a slightly larger number of advanced level students and a somewhat better score on chronological skills). Nevertheless, this group did not achieve a better result on the post-test than the other group. It was demonstrated during the post-test that the group presented with the imaginative-associative approach achieved a greater learning effect on virtually all points compared with the group confronted with the chronological approach. Therefore, the imaginative-associative aids worked better for an orientation in historical time compared to the chronological approach. This was not only the case where there was keyword recognition in the name of a time period and the description of an event, and it also occurred where acquired knowledge had to be applied to unknown events. The imaginative-associative time periods could serve as memory locations to allow events to be allocated to the correct time without keyword recognition or recognition based on numbered years. This applies to both reproduction as well as application, or the correct positioning in time of events not dealt with in the subject matter. Students found the imaginative-associative approach more comprehensible than the chronological approach. The effects described here will perhaps be stronger within regular education where students study their homework tasks seriously in order to prepare for tests.

It should be noted that the hypothesis stated at the outset of this chapter has been confirmed on two points, both in terms of the recall of information as well as the use of information as a reference framework. However, it is noticeable that the scores on the reproductive category were lower than those for the application category, which could imply that reproduction of information is enhanced less strongly by the imaginative-associative approach than the use of information as a reference framework.

On the basis of Friedman's (1992a, 2008) theory of "islands of time" and the research carried out by Levstik and Barton (2008) with elementary school students, it was to be expected that the imaginative-associative method would work better than the chronological approach. However, Friedman's work was focused on autobiographical memory. There was therefore some uncertainty as to whether these types of mechanisms would also occur when learning history, where there are other time dimensions and where there is no recall of memories of personal events. The experiment described in this chapter appears to have shown this. Barton and Levstik's (Barton, 2002) work exclusively involved the use of images and did not involve any events. The students locating images in time in Barton and Levstik's (Barton, 2002) study made use of a number of strategies that could not really have played a role in the experiment described here. The "progress" strategy ("what appears to be more like us should be placed in a more recent time period") was possibly the only one of these strategies used by a number of students, where they had to locate events that appeared to be modern, such as opening a shopping mall or building an airport, and decided that these must have occurred "now" or somewhere since the year 2000, which in the case of this experiment did not result in positive scores.

The experiment provides evidence that students in secondary schools within the age group 13 to 16 created memory locations during the imaginative-associative approach allowing them to more successfully link historical events with periods in comparison to those using numbered years and centuries. This concurs with the observation that orientation in time using chronological tools is relatively remote from the way in which people usually deal with time.

The design of the imaginative-associative eras for the history of Aruba as described in this experiment is of a similar kind as the ten eras proposed in 2001 in the report of the Dutch Committee on Historical and Social Education (CHMV, 2001; Wilschut, 2009). Figure 4.12 shows the ten eras that are presently used as a backbone of the Dutch history curriculum, and the six imaginative-associative eras that have been designed for the experiment described in this chapter. The resemblance between the two is no coincidence. When the CHMV proposed a curriculum based on ten associative eras, it was assumed that this would enable students to become more easily orientated in time and that the reference framework acquired would be more usable and enduring than the frameworks for teaching of history that had been used to date. However, there was no scientific basis for this assumption. This deficiency was acknowledged by the Commission. Both the CHMV, as well as its predecessor, the Dutch Committee on History Education (CG) highlighted the lack of research into history education and argued for an enhancement of the scientific basis for history education (CG, 1998; CHMV, 2001).

The history curriculum based on ten eras was introduced into the attainment targets for primary education, for the foundation years of secondary education, as well as into the history examination programs for secondary education in the Netherlands during the years 2005, 2006 and 2007. Since then the ten eras have become generally known to Dutch students. It is therefore no longer possible to measure the effect of the use of that curriculum without coming across the disruptive situation that one school may place a greater emphasis on it, or teach it differently than another school. Either way, researchers will always be faced with students with prior knowledge, and will never come across students starting with a blank slate. For that reason subject matter was chosen for this experiment from a totally un-

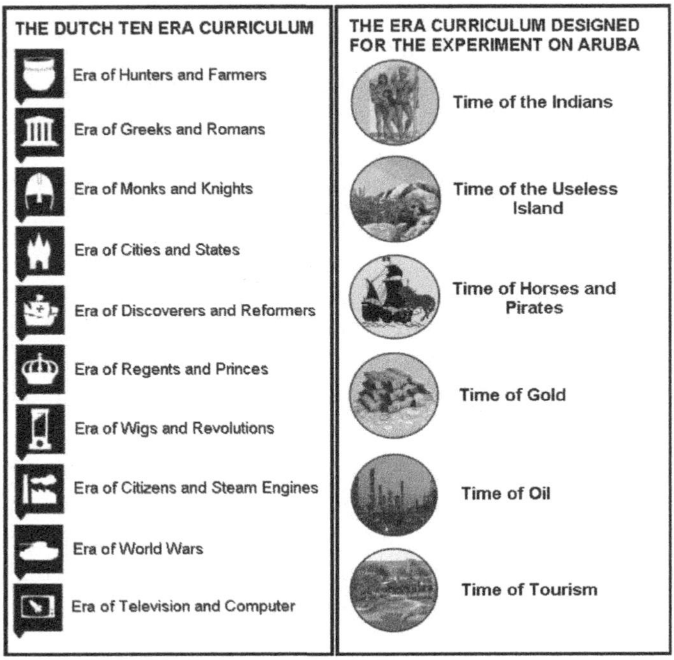

Figure 4.12

known context, and imaginative-associative eras were designed fit for that specific subject matter. One of the subsidiary aims of this experiment, however, was to test whether the assumptions of the CHMV in 2001 were, in fact, correct. The result of this experiment suggests that this was indeed the case.

5

Conclusion

> If history is a universal human interest...there are already as many historians as there are human beings, and the question is not "shall I be an historian or not?" but "how good an historian shall I be?" (Collingwood, 1930, p. 3)

> The best we can usually do is to forget about time, but life soon finds a way to remind us... (Grondin, 2008, p. xi)

It is impossible for humans not to have any notions about the past. People have them because human existence outside of time is inconceivable. Nobody could live without some thoughts about the past, the present and the future. Although people as a rule do not expend a great deal of conscious thought on the passage of time, the temporality of everything in existence is an inescapable fact. The question is not therefore whether the past deserves any attention, but rather how it should be given focus. A specific type of time consciousness was studied in this book: a historical consciousness of time. This form of consciousness is not one that comes easily to people. In Collingwood's words, one can either become good at it, or not. In terms of history teaching, not every approach to dealing with time is equal. The question in history teaching should be: how good shall we become at dealing with the past?

Images of Time, pages 183–191

The issue of which orientation on the past should be central to history education is often subject to fierce debate, not just among professionals, but also among the general public, politicians, the press, and opinion leaders. More than any other school subject, history seems to be everyone's possession. An important controversy, for instance, is the question of whether history teaching should concentrate in particular on presenting a more or less fixed body of knowledge—a cultural heritage or "canon"—or whether it should focus on specific historical attitudes, thinking, and ways of reasoning. It became clear in the first chapter of this book how history education through the course of its existence has wrestled with its core task: to attempt to break through the present day unilateral perspective and to take perspectives from the past seriously. It became evident that social and political imperatives normally interfere with the process of dealing properly with perspectives from other times. Historical consciousness of time not only puts into perspective and thereby makes relative today's self-evident insights and values, but also all too easy projections from the present to the past that are associated with the story of "how we have become who we are." It is questionable how identities in terms of communal, national or other group pasts could be created if a historical consciousness of time were to be taken seriously in this process. This explains the tension between the task faced by history teaching to develop a relativizing consciousness of historical time within students and the demands from a social and political reality, which are often set at enhancing a fixed national identity and shared values. It became clear in the first chapter that social and political demands have placed pressure on history teaching not only in recent years, but actually throughout its entire existence. This book has been partly written as a counterbalance to this pressure.

The passage of time is something that the average human experiences as self-evident and unproblematic. However, those who are consciously involved in the passage of time and images of time, such as historians, realize how impenetrable the phenomenon is. The apparent obviousness and, on second thoughts, great intangibility of time, defined in the oft cited response by St. Augustine to the question about the essence of time—"So, what then is time? When nobody asks me, I know the answer. However, when I try to explain it to someone who asks, I don't know what it is" (St. Augustine, 1992, Book XI, 14:17)—provides a specific context for history teaching. Students probably bring a number of obvious perceptions about the passage of time along to the classroom, which may only partly apply to a historical consciousness of time or not at all. In educational jargon, these are referred to as "misconceptions." It became clear in the second chapter of this book how a historical consciousness of time on the one hand responds to objectivating chronological standards and on the other to subjectivating imagery. There are "rules" applying to both aspects that need to be learned—rules that sometimes contradict spontaneous intuitions. For instance, an objective chronological rule is that the number of years determines the "actual length" of a time period, and that, for instance, the Middle Ages lasted longer than the distance of time that separates us from the Middle Ages. "Medieval conditions" may relate to a diversity of a thousand years; this is somewhat different to the image people normally have when referring to this expression.

One of the rules from the subjectivating domain is that a historical consciousness of time demands that one attempts to leave one's own position in time—"temporal decentering"—which conflicts with strong intuitions to the contrary. Intuitive awareness of time, as became apparent in the second chapter, often consists of components of daily/cyclical time awareness, social time awareness and an awareness of a distant, "mythical" past. This could be contrasted with disciplinary rules of linear chronology and periodization, of dealing with generations and relics in a historical manner, as well as with anachronism and contingency. The historian contrasts the eternals truths of the myth with the relativism and uncertainty of a differentiated, unpredictable and fortuitous image of the passage of time. The six key concepts of a historical consciousness of time are bound to discipline-specific rules that have to be *learned,* as they do not come naturally to people.

How this learning is taking shape was explored in the third chapter. Historical consciousness of time is a topic that as such has not been studied by itself for a long time. Consciousness of time was summarized as a single category: more or less obvious, daily, natural time. The fact that children had difficulties dealing with this was known, but it was thought that this was only a problem affecting primary school children. It was thought that after this age, children could easily get to grips with the problem. This position had two consequences. On the one hand people were of the opinion that teaching history to young children was pointless, given that they had no awareness of time. On the other hand the specifics of a historical consciousness of time were barely acknowledged and it was assumed that once children had developed a reasonable understanding of time, they would be able to create a historical image of the past without too many problems. Both assumptions appeared to be incorrect once the historical consciousness of time was placed into a distinct, separate category in research: on the one hand children appeared to be able to create images of former times much earlier than previously thought, and on the other it was apparent that there were more and different types of difficulties associated with developing an adult historical consciousness of time than had been realized.

History teaching that is explicitly based on the development of a historical consciousness of time—as developed in Chapter 2 of this book around the six key concepts of chronology, periodization, generations, relics, anachronism and contingency—has not been developed up to now. This may be due to the fact that other demands are so often imposed on history teaching rather than the developing of such a time awareness, as became apparent in Chapter 1. This does not detract from the fact that a significant proportion of research into history teaching can be interpreted in relation to the six key concepts. Research has been particularly aimed at what students tend to do spontaneously (i.e., their misconceptions) and has not to date been very concerned with developing educational strategies to deal with these misconceptions. The description of the historical consciousness of time in six key concepts also highlighted important gaps in the research into students' (mis)conceptions, particularly around the concepts of chronology, periodization, and the historical generation concept: the relationship between a historical and a social past as expressed in the use of an "us-" and "them-perspective." It may, however, be concluded on the basis of the research that expertise in teaching and learning history still has to travel a long way in terms of developing educational approaches to deal with the many misconceptions held by students regarding anachronism (particularly presentism, the use of modern standards) and contingency (particularly knowledge of outcomes and evaluating historical developments in a straight line with the present day as a logical outcome). A research agenda for teaching and learning history was presented at the end of Chapter 3.

A modest attempt was made in Chapter 4 to start to implement this research agenda. It was decided to further explore those questions that—on the basis of the data from the previous chapters—would generate a clear hypothesis. Chapter 4 was particularly concerned with a question that is central to history teaching, and one that sorely tasks many students in their homework: how should the use of numbered years and timelines be evaluated as anchor points in learning about historical time? The assumption was, on the basis of psychological and educational research, that the value of numbered years and timelines could be seen as being less significant than is commonly felt in history teaching. This supposition has been confirmed by empirical research. Students confronted with associative time periods and stories appeared to be better able to generate a historical reference framework about an unfamiliar period of history than those students presented with numbered years and centuries. This did not just apply for knowledge reproduction, but also for knowledge application: students placing unknown events in the correct time. And it did not just apply to events that could be directly associated with the name of time period: it also applied to other events that could only be linked to a time period through indirect associations. The time periods had been deliberately named using the type of nomenclature used for the ten eras in the history curriculum as proposed by the Dutch Committee on Historical and Social Education in 2001 and introduced in the Netherlands some years ago. There had, however, been a lack of a scientific basis for that curriculum. The results of the research reported in Chapter 4 seem to provide this scientific basis.

The research described in Chapter 4 is no more than a start to implementing the research agenda set out in the study described in Chapters 2 and 3. Responding to the numerous questions raised—not just the questions formulated at the end of Chapter 2, but also those questions arising from the collection of psychological insights, and insights from history theory and pedagogy—will require a long series of research into teaching and learning history.

The most important contributions this book has provided to the development of expertise on teaching and learning history may be summarized in the following points:

- An operational definition has been developed for a "historical consciousness of time" for use in history teaching. This is based around six key concepts: chronology, periodization, generations, relics, anachronism, and contingency.
- This historical consciousness of time has been contrasted against alternatives in the form of daily/cyclical, social and mythical awareness of time. It has been suggested that the key concepts of chronology and periodization are counterweights to daily/cyclical time awareness, the key concepts of generations and relics are counterweights to social time awareness, and the key concepts of anachronism and contingency are counterweights to mythical time awareness. These combinations may be important to the development of expertise on teaching and learning history as students probably bring along their daily/cyclical, social and mythical time awareness to the classroom as obvious intuitions, to which history teaching based on a historical consciousness of time could be used as a counterbalance.
- A relationship was highlighted between the results of psychological research into autobiographical memory and educational research into learning a historical consciousness of time. It was suggested—on the basis of existing research results, as well as on the basis of the empirical study described in Chapter 4—that the location theory of memory, according to which people make use

of associative "time islands" in their memories, also has a validity for learning history.

- It became clear on the basis of empirical research that numbered years and centuries (mathematical chronology) do not play such an important role in terms of getting to grips with the chronology of an unfamiliar period of history and that the use of associative contexts should be preferred in this process. This has created a scientific basis for the use of an era curriculum as proposed by the Dutch Committee on Historical and Social Education in 2001 and introduced into Dutch history education.

Consequences for Teaching

Any subject-specific educational study may be expected to generate useful information for the teaching of the subject concerned. Unfortunately, educational research is not advanced far enough to provide guidelines for good education that would be able to guarantee success to any significant extent. This study has provided a modest contribution to progressing research into history teaching. Therefore only general conclusions may be drawn for educational practice.

Advice may be provided with some certainty on the basis of the empirical study described in Chapter 4. It follows from this that memory locations need to be created to facilitate an understanding of time and that these need to be formed as associative contexts within which students are able to position historical data. Memorizing numbered years and working with timelines probably has little use if these are not utilized alongside narratives and imaginative contexts (including the use of lots of images). Dates and numbered years do not correspond to the way people "naturally" form an image of the passage of time. The use of calendars and chronology is an artificial addition to the experience of time, one which we find difficult to deal with. Numbered years therefore do not deserve a significant place in history teaching as anchor points.

The second recommendation is not based on the research presented here, but on that from others. However, it does involve to a significant extent the use of imaginative memory locations, which is why it is included here. Teaching about a single time period by itself means it becomes superfluous for students to compare different time periods and to actually use their diverse memory locations. This means students do not easily develop a historical reference framework in this way. The advice therefore is to remove the highly structured, chronological approach to teaching and to present different time periods numerous times, in comparative studies, in different orders and different combinations. The chronologically structured approach to teaching needs—somewhat paradoxically—to be removed from teaching into order to promote chronological understanding. Those who do not travel through time but remain in a single time period do not develop any cognitive methods for viewing time periods together and comparing different time periods. It is exactly this, the comparison of different time perspectives, that is the basis for a historical consciousness of time.

The third recommendation is relatively practical and reasonably easy to follow. Attempts should be made in history teaching to systematically avoid the use of phrases such as "not yet" and "already" (with implicit reference to the present as a standard), and always to use the "them-perspective" and never refer to "us" when talking about the historical past. This could not only be applied by history teachers in their own use of language, but could also be systematically taught to students (including highlighting "errors" in the use of contested expressions in tests and essays, or even in verbal contributions in class). This, of course, would make most sense if the students also understand why they should not use these expressions.

A lot of effort is made in history education to teach students the correct facts about various cultures from the past. The results of the overview in Chapter 3 show, however, that students are rarely taught how they should think about and evaluate these facts. Is it correct to evaluate the past using today's standards? And if we refrain from this, are we then not simply condoning all of the massacres, slavery and misery from the past? Is it true that "we" have our own national history that has made "us" what we are today? Is the course of history characterized by progress? Do coincidences and unintended consequences govern the course of history? Is the present the logical outcome of preceding ages? It is likely that every history teacher will deal with one or more of these issues at some time. However, there is also a significant chance that this will be incorporated in the background as a sideline to the "real" subject matter, dealing with, for example, the Romans or feudalism or the Russian revolution. Should they not, however, be the principal elements in the subject matter? Would it not lead to a more persistent and significant result of history education if students were to learn to naturally display patterns of thought belonging to a historical consciousness of time, rather than remembering when Caesar crossed the Rubicon? This question should be related to the fact that students are highly likely to forget memorized facts. However, once learned, a particular thinking attitude is unlikely to rapidly disappear.

Significance for Society

This takes us back to the social role of history education, which was the starting point for this book. The systematic development of particular ways of thinking that are associated with a historical consciousness of time may have significant implications for society. The first condition for this are methods of thinking around the concepts of chronology and periodization, which are not a goal in themselves: those who wish to develop a differentiated and nuanced image of the past first have to have an insight into the various periods of history, how these have to be placed in order, and which characteristics belong to each of them.

A systematic removal of modern-day standards and not falling back on one's "own" past through the use of the "us-perspective" may serve to promote insights into and respect for others. This may be of use to a society where diversity is an important phenomenon. One's own position—not only in the sense of opinion, but also in the sense of one's location, if this is used in the sense of location in time—is important here, as well as other perspectives that may occasionally be radically different to one's own, so different that the first reaction might be that something bizarre or insane is going on. Learning to understand that the past has not followed a direct and logical path that has naturally led to the present may serve to relativize the present, to understand that something completely different might also have occurred, and that this is also a possibility for the future. This may lead to the insight that one's own opinions, cultural norms, values, political and religious convictions are—for the main part—the result of coincidental developments, which may help to avoid viewing one's own perspective as being absolute, and thus to be more tolerant of other's—equally coincidental—perspectives, and thereby, for instance, prevent or minimize the effects of social conflict.

Learning to understand that historical developments often included unintended consequences may contribute to less of a conviction of the manipulability of society, less conviction about the greatest, sometimes totalitarian ideological perspectives of the future "we deserve" and a greater belief in an open, liberal and democratic society characterized by "piecemeal social engineering" (Popper, 1961, pp. 64–70).

All these insights and ways of thinking may contribute to citizenship in a democratic society. In the 1970s Kocka (1977) pointed out that this type of history would only come about in open democratic societies, and would also serve as a foundation for societies such as these. History cannot provide such a foundation by conveying the message of a national heritage and the construction of a national identity rooted in shared, unchangeable values and norms. A historical consciousness of time provides such a foundation in a far more profound and significant manner.

Notes

Introduction

1. In many European languages there is a word that looks like the English "didactic"—(German) Didaktik, (French) didactique, (Spanish) didactica, (Italian) didattica—which refers to expertise in the field of teaching and learning. It does not exactly correspond with the English "didactics"—which has a different connotation and does not refer to the academic discipline which is discussed in this book—and therefore in this translation the use of "didactics" has been avoided. In German and Dutch there are also the words "fachdidaktik" or "vakdidactiek" (subject-educational science) and "geschichtsdidaktik" or "geschiedenisdidactiek" (history-educational science). Because this introduction deals with the position of this discipline in the academic spectrum, these words are used frequently. Because there is no good English equivalent of them, the word "vakdidactiek" (equivalent to German "fachdidaktik") has been translated as "domain-specific educational science," and the word "geschiedenisdidactiek" (equivalent to German "geschichtsdidaktik") as "expertise in teaching and learning history." The word "methodology," which is frequently used in English, has not been chosen because "didactiek" is not only concerned with teaching methods, but also deeply involved in (the structure) of subject matter and modes of thought.

Chapter 1

1. The best known of these historical images for schools about Dutch history were created by Johan Herman Isings: 44 plates produced during the period 1911–1970. The fact that 29 of the 44 plates date from after the Second World War illustrates how well the nineteen fifties and sixties fitted in with the prewar tradition. Refer to Niemeijer (2000).

Images of Time, pages 193–196

2. For instance, examples were included in the chapter on teaching objectives about "Hungary 1956" where students had to determine what had actually happened on the basis of two eyewitness accounts—one from a Russian and one from an American (Dalhuisen, Geurts, & Toebes, 1976, pp. 49–56). This lesson was a literal translation of Fenton's 1966 (pp. 159–164 and 182–183) "How the Historian Decides What Is Fact."
3. The notion of time-dependency was made explicit in the educational objectives developed by the HEG alongside other aspects of "dependency": "To appreciate the motives, interests, value systems and expectations of other people, *taking into consideration that those people, unlike us, were not able to acquire post hoc knowledge of later events*... To evaluate situations and people from the past, *taking consideration of the beliefs of that time*" (HEG, 1986, p. 17, italics added).
4. The series of schoolbooks *Sprekend Verleden* (*The Past Speaks,* published by Nijgh & Van Ditmar and subsequent publishers from 1971 onwards) developed by Dalhuisen is noticeable for its traditional choice of subject matter, just like the contents of much of the English SCHP curriculum.
5. For example, a survey was published in *Kleio* (Dutch magazine for history teachers) in 1999 to list the names and facts that should be included in the canon of chronological knowledge that the commission should produce (Van Boxtel, 1999a; Van Boxtel, 1999b).
6. Examples from the headlines of *De Volkskrant* of February 8, 2001: *"Feiten weer basis lesstof geschiedenis"* (History lessons to be based on facts again) or: *"Op naar de tastbare geschiedenis"* (Towards a more tangible history), *NRC-Handelsblad,* February 22, 2001.
7. It should be noted, however, that national ideals were also a significant factor in the historical sciences of the nineteenth and early twentieth centuries.

Chapter 2

1. Hall in his work *Beyond Culture* (1976) makes a distinction between "high context culture" and "low context culture." The first develops within groups with a strongly shared experience where little is made explicit and much has to be drawn from the context. Objective time does not belong to this world, as it is made explicit.
2. The movement of the moon around the earth (the sidereal month) takes 27.321661 days and a full cycle of phases of the moon (the synodic month) takes 29.530588 days. The Gregorian calendar differs from the sidereal and synodic months in order to allow the lunar year to coincide with the solar year. The Islamic calendar, however, is based on the lunar year. Encyclopedia Britannica: "month."
3. By "awkward" I mean that it is not handy for calculations that an hour has 60 minutes, and not (for instance) a hundred, and that a day comprises 24 hours and not (for instance) 10 hours. If a day consisted of 10 hours of a hundred minutes then one minute "new style" would amount to 86.4 (current) seconds. Hours and minutes could then be decimalized, just like dollars and cents, and meters and centimeters. The division into 24 hours and 60 minutes originates from ancient Mesopotamia, and as far as I know there are no other

measurement systems from this culture currently still in use. The division of the year into twelve months of 30, 31 and 28 days is not very handy either.

4. The laws are as follows: 1) An object that is not subject to any force will remain stationary or move in a straight line at the same speed. 2) The acceleration of an object is the function of its mass and the force applied to it, in such a way that force = mass × acceleration. Establishing "the same speed" requires the same distance to be covered in equal time intervals. Measuring "acceleration" requires that constantly increasing distances are covered in equal time intervals. The fixed relationship between time intervals and other natural instances (mass, force) defines natural time, as established first by the mathematician Leonhard Euler (1707–1783) in *Réflexions sur l'espace et le temps,* 1748.
5. Roeckelein is of the opinion that Newton's treatment of time was more theological than scientific (Roeckelein, 2008, p. 24).
6. Newton, *Principia Mathematica Philosophiae Naturalis,* Definitiones, Scholium I (p. 11): "Tempus absolutum verum et mathematicum, in se et natura sua absque relatione ad externum quodvis, aequabiliter fluit, alioque nomine dicitur duratio."
7. This counterargument against a temporal vacuum appears in Leibniz's *Nouveaux Essais sur l'Entendement Humain* (1704) and is cited by Le Poidevin (2003, p. 22).
8. Quid autem familiarius et notius in loquendo commemoramus quam tempus?
9. Et tamen dicimus longum tempus et breve tempus, neque hoc nisi de praeterito aut futuro dicimus. Et tamen, domine, sentimus intervalla temporum et comparamus sibimet et dicimus alia longiora et alia breviora. Et metimur tamen tempora.
10. Tempora sunt tria, praesens de praeteritis, praesens de praesentibus, praesens de futuris.
11. Sunt enim haec in anima tria quaedam et alibi ea non video, praesens de praeteritis memoria, praesens de praesentibus contuitus, praesens de futuris expectatio.
12. Inde mihi visum est nihil esse aliud tempus quam distentionem; sed cuius rei, nescio, et mirum, si non ipsius animi.
13. Nam quod eius iam peractum est, utique sonuit, quod autem restat, sonabit atque ita peragitur, dum praesens intentio futurum in praeteritum traicit: *A part of it [the sound] has already been completed in the way it has sounded, yet the remainder will sound and thus become completed, while the present 'intentio' transfers the future into the past.*
14. Ricoeur comments that the line being set out here was pursued again in the twentieth century by thinkers such as Husserl and Heidegger (Ricoeur, 1984, p. 14).
15. Husserl records his debt to St. Augustine in the first sentences of his *Vorlesungen* and adds modestly to this: *"Denn herrlich weit gebracht und erheblich weiter gebracht als dieser grosse und ernst ringende Denker hat es die wissenstolze Neuzeit in diesen Dingen nicht"* (Because New Times boasting their knowledge have not taken things very far or much farther than this great and seriously wrestling thinker) (1928, p. 2). Similar to Bergson and St. Augustine, Husserl uses the

example of experiencing music and sound in order to explain temporal consciousness.

16. It is noticeable in relation to this that the anthropologist Hall did not include historical time in his system.
17. "Jede Epoche ist unmittelbar zu Gott, und der Wert beruht gar nicht auf dem, was aus ihr hervorgeht, sondern in ihrer Existenz selbst, in ihrem eigenen Selbst."
18. "Hermeneutics" may be understood as the study of *verstehen* and interpretation.
19. The cited works are: Dilthey, *Über das Studium der Geschichte, der Wissenschaften vom Menschen, der Gesellschaft und dem Staat* [About the Study of History, the Sciences of Man, Society and the State] (1875), and Pinder, *Das Problem der Generationen in der Kunstgeschichte Europas* [The Problem of Generations in the Art History of Europe] (1926).
20. Carr refers to a "historical past," also where there is a "pre-thematic interest," while Oakeshott explicitly reserves the term "historical interest" for the image of the past resulting from an intended aim of the study. It appears to me that the distinction used by Oakeshott is the clearest and Carr's "pre-thematic interest" should be allocated to the "social past."
21. "Man hat der Historie das Amt, der Vergangenheit zu richten, die Mitwelt zum Nutzen zukünftiger Jahre zu belehren, beigemessen. So hoher Ämter unterwindet sich gegenwärtiger Versuch nicht; er will bloss zeigen, wie es eigentlich gewesen."
22. Van der Dussen (1988, p. 43) cites a speech by the French historian Fustel de Coulanges, who reacted to receiving applause on his lecture from his students by saying, *"Gentlemen, you should not applaud. It is not I who am speaking here, but history is speaking through me."*
23. The moment of creation coincides with August 11, 3113 BC in the Western calendar.
24. The cycle of 144,000 days is the result of 20 times 18 times 20 times 20 days (the shorter, underlying cycles).

Chapter 3

1. Stow and Haydn even based some of their conclusions on this still in 2000 (Stow & Haydn, 2000, pp. 87–90).
2. Grever was of the same opinion in 2009, and also adopted the classic viewpoint of the *development* of temporal awareness, rather than speaking about a *learning process.* (Grever, 2009, p. 442).
3. An example is provided in Chapter 4 in Lévesque's book (2008, pp. 62–86), where the title suggests a focus on continuity and change, although it is largely devoted to causality.

References

Aldrich, R. (2006). *Lessons from history of education.* New York, NY: Routledge.

Alter, P., Rohlfes, J., Wolfrum, E., & Wolle, S. (2006). *Deutsche Geschichte. Wie wir wurden, was wir sind. 20. Jahrhundert* [German History. How we became, what we are, 20th century]. Stuttgart, Germany: Klett Verlag.

Amsing, H. (2002). *Bakens verzetten in het voortgezet onderwijs 1863–1920. Gymnasium, h.b.s. en m.m.s. in onderwijssysteem, leerplan en geschiedenisonderwijs* [Changing directions in secondary education 1863–1920: Gymnasium, h.b.s and m.m.s in educational system, curriculum and history teaching]. Delft, The Netherlands: Eburon.

Anderson, B. (1991). *Imagined communities: Reflections on the origin and spread of nationalism* (2nd rev. ed.). New York, NY: Verso.

Angvik, M., & von Borries, B. (Eds.). (1997). *Youth and history: A comparative European survey on historical consciousness and political attitudes among adolescents. Vol. A: Description. Vol. B: Documentation.* Hamburg, Germany: Körber-Stiftung.

Ankersmit, F. R. (1983). *Narrative logic: A semantic analysis of the historian's language.* The Hague, The Netherlands: Martinus Nijhoff.

Ankersmit, F. R. (1984). *Denken over geschiedenis. Een overzicht van moderne geschiedfilosofische opvattingen* [Thinking about history: A survey of modern views on the theory of history]. Groningen, The Netherlands: Wolters-Noordhoff.

Ankersmit F. R. (1989). Over geschiedenis en tijd [About History and Time]. *Groniek 103/104,* 11–26.

Ankersmit, F. R. (2007), *De sublieme historische ervaring.* Groningen, The Netherlands: Historische Uitgeverij. [English edition: Ankersmit, F. (2005). *Sublime historical experience.* Stanford, CA: Stanford University Press. References are to the Dutch edition.]

Images of Time, pages 197–216

Ashby, R., & Lee, P. (1987). Children's concepts of empathy and understanding in history. In C. Portal (Ed.), *The history curriculum for teachers* (pp. 62–88). London, UK: Falmer Press.

Aristotle. (2001). *Physics.* Book IV, 10-14. P. Gerbrandy (Trans.). *De mens is een dier dat kan denken. Bloemlezing uit de Griekse en Romeinse filosofie.* Amsterdam: Uitgeverij Contact.

Aristotle. (1932). *Poetics.* Aristotle in 23 Volumes, Vol. 23, W.H. Fyfe (Trans.). Cambridge, MA: Harvard University Press. Retrieved from http://www.perseus.tufts.edu/cgi-bin/ptext?lookup=Aristot.+Poet.+1450a

Augustine. (1992). *Confessions.* J. O'Donnell (Trans.). Stoa Corporation: Retrieved from http://www.stoa.org/hippo/

Aveni, A. F. (2002). *Empires of time: Calendars, clocks and cultures* (rev. ed.). Boulder, CO: University Press of Colorado.

Baker, E. (1994). Learning-based assessments of history understanding. *Educational Psychologist,* 29(2), 97–106.

Barton, K. C. (2001). A sociocultural perspective on children's understanding of historical change: Comparative findings from Northern Ireland and the United States. *American Educational Research Journal, 38*(4), 881–913.

Barton, K. C. (2002). "Oh, that's a tricky piece!": Children, mediated action and the tools of historical time. *The Elementary School Journal, 103*(2), 161–185.

Barton, K. C. (2004). Mary G. Kelty: The most important social educator no one has heard of? In L.M. Burlbaw & S.L. Field (Eds.), *Explorations in curriculum history* (pp. 161–184). Charlotte, NC: Information Age Publishers.

Barton, K. C. (2008). Narrative simplifications in elementary students' historical thinking. In L.S. Levstik & K.C. Barton (Eds.), *Researching history education: Theory, method and context* (pp. 183–208). New York, NY: Routledge.

Barton, K. C. (2009). The denial of desire: How to make history education meaningless. In L. Symcox & A. Wilschut (Eds.), *National history standards: The problem of the canon and the future of teaching history* (pp. 265–282). Charlotte, NC: Information Age Publishers.

Barton, K. C., & Levstik, L. S. (1996). "Back when God was around and everything": Elementary children's understanding of historical time. *American Educational Research Journal, 33*(2), 419–454.

Barton, K. C., & Levstik, L. S. (2008). "It wasn't a good part of history": National identity and students' explanations of historical significance. In L. S. Levstik & K. C. Barton (Eds.), *Researching history education: Theory, method and context* (pp. 240–272). New York, NY: Routledge.

Bauvois-Cauchepin, J. (2002). *Enseignement de l'histoire et mythologie nationale. Allemagne-France de début de XXe siècle aux années 1950* [History teaching and national mythology. Germany and France from the beginning of the 20th century to the 1950s]. Bern, Switzerland: Peter Lang.

Benveniste, E. (1974). *Problèmes de linguistique générale* [Problems of general linguistics], Vol II. Paris: Gallimard.

Bergmann, K. (1976). Warum sollen Schüler Geschichte lernen? [Why should students learn history?], *Geschichtsdidaktik, Probleme, Projekte, Perspektiven, 1*(1), 3–14.

Bergson, H. (1888). *Essai sur les données immédiates de la conscience* [Essay on the direct intuitions of conscience]. Electronic edition by Jean-Marie Tremblay, Chicoutimi, Québec. Retrieved from http://classiques.uqac.ca/classiques/bergson_henri/essai_conscience_immediate/essai_conscience.pdf

Bianchi, S. (1982). *La révolution culturelle de l'an II* [The cultural revolution of the year II], Paris, France: Éditions Aubier.

Blaas, P. B. M. (1988). *Anachronisme en historisch besef. Momenten uit de ontwikkeling van het Europees Historisch Bewustzijn* [Anachronism and historical awareness: Moments in the development of European historical consciousness]. Den Haag, The Netherlands: Nijgh en Van Ditmar Universitair.

Blaas, P. B. M. (2001). Vorm geven aan de tijd. Over periodiseren [Giving shape to time: About periodization]. In M. Grever & H. Jansen (Eds.), *De ongrijpbare tijd* (pp. 35–48). Hilversum, The Netherlands: Verloren.

Block, R., & Zakay, D. (2008). Timing and remembering the past, the present and the future. In S. Grondin (Ed.), *Psychology of time* (pp. 367–394). Bingley, UK: Emerald.

Blom, J. C. H. (1986), 'In de ban van goed en fout? Wetenschappelijke geschiedschrijving over de bezettingstijd in Nederland' [Under the Spell of Good and Evil? Scientific Historiography on the German Occupation of the Netherlands], in: Abma, G., Kuiper, Y. & Rypkema, J. (eds.), *Tussen goed en fout. Nieuwe gezichtspunten in de geschiedschrijving 1940–1945* (pp. 30–52). Franeker, The Netherlands: Wever.

Blom, J. C. H. (1989). De Tweede Wereldoorlog en de Nederlandse samenleving: continuïteit en verandering [The Second World War and Dutch society: Continuity and change]. In J. C. H. Blom (Ed.), *Crisis, bezetting en herstel. Tien studies over Nederland 1930–1950* (pp. 164–183). Den Haag, The Netherlands: Nijgh & Van Ditmar Universitair.

Blom, J. C. H. (2010). *Hoe was het mogelijk? De Holocaust in de context van de Tweede Wereldoorlog* [How was it Possible? The Holocaust in the Context of the Second World War]. Leiden, The Netherlands: Universiteit Leiden.

Bloom, B., & Krathwohl, D. R. (Eds.) (1956). *Taxonomy of educational objectives: The classification of educational goals, by a committee of college and university examiners.* New York, NY: Longmans/Green.

Boekholt, P. Th. F. M., & de Booy, E. P. (1987). *Geschiedenis van de school in Nederland* [History of the school in the Netherlands]. Assen, The Netherlands: Van Gorcum.

Boot, E. (1989). Tijd zonder grenzen? Maya tijdrekening in heden en verleden [Time Without Limits? Maya chronology in the present and the past]. *Groniek, 103/104,* 165–177.

Booth, M. (1994). Cognition in history: A British perspective. *Educational Psychologist, 29*(2), 61–69.

Bradley, N. C. (1948). The growth of the knowledge of time in children of school-age. *British Journal of Psychology, 38*, 67–78.

Bransford J. D., Brown, A. L., & Cocking, R. R. (Eds.). (1999). *How people learn: Brain, mind, experience, and school.* Washington, DC: National Academies Press.

Braudel, F. (1966). *La Méditerranée et le monde méditerranéen à l'époque de Philippe II* [The Mediterranean and the mediterranean world in the times of Philip II] (2nd ed.). Paris, France: Colin.

Braudel, F. (1979). De lange duur [The long duration]. In F. Braudel (Ed.), *Geschiedschrijving* (pp. 50–84). Baarn, The Netherlands: Ambo.

Broad, C. D. (1983). *An examination of McTaggart's philosophy* (Vol. II, part 1). Cambridge, UK: Cambridge University Press.

Bruner, J. S. (1960). *The process of education.* Cambridge, MA: Harvard University Press.

Bryman, A. (2008). *Social research methods* (3rd ed.). Oxford, UK: Oxford University Press.

Burke, P. (2002). Western historical thinking in a global perspective—10 theses. In J. Rüsen (Ed.), *Western Historical Thinking. An Intercultural Debate* (pp. 15–30). New York, NY: Berghahn Books.

Butterfield, H. (1973). *The Whig interpretation of history.* Harmondsworth, UK: Penguin Books.

CEGES (Commissie Eindexamenexperimenten Geschiedenis en Staatrinrichting) (1979). *Eindrapport Commissie Eindexamenexperimenten Geschiedenis en Staatsinrichting* [Final Report of the Commission on Experimental Exams in History and Polity]. The Hague, The Netherlands: Department of Education and Sciences.

CHMV (Commissie Historische en Maatschappelijke Vorming) (2001). *Verleden, heden en toekomst* [*Past, present and future*]. Enschede, The Netherlands: Stichting Leerplan Ontwikkeling.

Carr, D. (1986). *Time, narrative and history.* Bloomington, IN: Indiana University Press.

Carter, J. A. (2003). Telling times: History, employment and truth. *History and Theory, 42*(1), 1–27.

Committee on History Education (CG). (1998). *Het verleden in the toekomst* [*The past in the future*]. The Hague, The Netherlands: Department of Education and Sciences.

Cohen, L., Manion, L., & Morrison, K. (2007). *Research methods in education* (6th ed.). New York, NY: Routledge.

Collingwood, R. G. (1930). *The philosophy of history.* Historical Association Leaflet # 79, London, UK: Bell.

Collingwood, R. G. (1973). *The idea of history.* Oxford, UK: Oxford University Press.

Coltham, J. B., & Fines J. (1971). *Educational objectives for the study of history: A suggested framework.* London, UK: The Historical Association.

Crawford, K. (1995). The battle for control of national curriculum history 1989–1994. *British Journal of Educational Studies, 43*(4), 433–456.

Daal, L. H. (1992). *Antilliaans verhaal: Geschiedenis van Aruba, Bonaire, Curaçao, Saba, St.-Eustatius en St.-Maarten* [Antillian Story: A history of Aruba, Bonaire, Curaçao, Saba, St. Eustatius and St. Maarten]. Zutphen, The Netherlands: De Walburg Pers.

Dalhuisen, L. G., & Korevaar, C. W. (1971). *De methode van onderzoek in het geschiedenisonderwijs* [The method of inquiry in history education]. The Hague, The Netherlands: Van Goor.

Dalhuisen, L. G., & Van der Dussen, J. W. (1975). *Wat is geschiedenis?* [What is history?]. Leiden, The Netherlands: Historisch-Didactisch Documentatiecentrum.

Dalhuisen, L. G., Geurts, P. A. M, & Toebes, J. G. (Eds.). (1976). *Geschiedenis op school* [History at school]. Groningen, The Netherlands: Wolters-Noordhoff.

Dalhuisen, L. G., Toebes J. G., & Verhagen. D. H. (Eds.). (1982). *Geschiedenis op school, 2 dln.*[History at school, 2 Vols.] Groningen, The Netherlands: Wolters-Noordhoff.

Dalhuisen, L., Donk, R., Hoefte, R., & Steegh, F. (Eds.). (2009). *Geschiedenis van de Antillen* [History of the Antilles]. Zutphen, The Netherlands: De Walburg Pers.

Dalhuisen, L., & Walhain, J. (1997). *Geschiedenis. Een vakvertaling bij Onderwijskunde* [History. A domain-specific translation of Onderwijskunde (Educational Science)]. Groningen, The Netherlands: Wolters-Noordhoff.

Davis, O. L., Yeager, E. A., & Foster, S. J. (Eds.). (2001). *Historical empathy and perspective taking in the social studies.* Lanham, MD: Rowman & Littlefield.

Dawson, I. (2009). "What time does the tune start?" From thinking about 'sense of period' to modelling history at Key Stage 3. *Teaching History, 135*, 50–57.

De Certeau, M. (1988). *The writing of history.* New York, NY: Columbia University Press.

Department of Education and Science. (1985). *General certificate of secondary education, the national criteria, history.* London, UK: Her Majesty's Stationary Office.

De Fouw, A., van Bennekom, J. A., & van de Bunt, A. W. (1950). *Geschiedenis van de Nederlanden* [History of the Netherlands] (vol. 3). Groningen/Djakarta: J.B. Wolters.

De Jong, G. (1977). *Kind, school en geschiedenis: Didactische aanwijzingen voor het geschiedenisonderwijs op de basisschool* [Child, school and history. Educational directives for teaching history at the primary school]. (7th ed.). Amsterdam, The Netherlands: Elsevier.

Donovan, M. S., Bransford, J. D., & Pellegrino, J. W. (Eds.). (1999). *How people learn: Bridging research and practice.* Washington, DC: National Academies Press.

Donovan, M. S. & Bransford, J. D. (Eds.). (2005). *How students learn history, mathematics and science in the classroom.* Washington, DC: National Academies Press.

Dostal, R. J. (1993). Time and phenomenology in Husserl and Heidegger. In C. B. Guignon (Ed.), *The Cambridge companion to Heidegger* (pp. 141–169). Cambridge, UK: Cambridge University Press.

Draaisma, D. (2006). *Waarom het leven sneller gaat als je ouder wordt. De geheimen van het geheugen* [Why life goes faster as we grow older: The secrets of human memory] (5th ed.). Amsterdam, The Netherlands: Rainbow Paperbacks.

Ebeling, H. (1973). *Didaktik und methodik eines kind-, sach- und zeitgemäßen geschichtsunterrichts* [Methodology of history education directed at the child, the subject matter and the present] (5th ed.). Hannover, Germany: Hermann Schroedel Verlag.

Ebels-Hoving, B. (1989). Het nut en nadeel der periodisering voor de geschiedenis [The utility and liability of periodization for history]. *Groniek, 103/104,* 115–132.

Eliade, M. (1991). *The myth of the eternal return, or cosmos and history* (9th paperback printing). Princeton, NJ: Princeton University Press.

Evans, R. J. (1997). *In defence of history.* London, UK: Granta Books.

Evans-Pritchard, E. E. (1939). Nuer time-reckoning. *Africa: Journal of the International African Institute, 12*(2), 189–216.

Fabian, J. (2002). *Time and the other: How anthropology makes its object.* New York, NY: Columbia University Press.

Fabian, J. (2005). If it is time, can it be mapped? *History and Theory, 44*(1), 113–120.

Feldman, D. H. (2004). Piaget's stages: The unfinished symphony of cognitive development. *New Ideas in Psychology, 22,* 175–231.

Fenton, E. (1966). *Teaching the new social studies in secondary schools.* New York, NY: Holt, Rinehart & Winston.

Fenton, E. (1967). *The new social studies.* New York, NY: Holt, Rinehart & Winston.

Fenton, E., & Good, J. M. (1965). Project social studies: A progress report. *Social Education, 29*(4), 206–208.

Fischhoff, B. (1975). Hindsight ≠ foresight: The effect of outcome knowledge on judgment under uncertainty. *Journal of Experimental Psychology, Human Perception and Performance, 1*(3), 288–299.

Fontaine, P. F. M. (1982). Basisgegevens voor de beginnende leraar [Fundamental data for the starting teacher]. In L. G. Dalhuisen, J. G. Toebes, & D.H. Verhagen (Eds.), *Geschiedenis op school* (vol. 1, pp. 401–420). Groningen, The Netherlands: Wolters-Noordhoff.

Foster, S. J. (2001). Historical empathy in theory and practice. In O. L. Davis, E. A. Yeager, & S. J. Foster (Eds.). (2001). *Historical empathy and perspective taking in the social studies* (pp. 167–181). Lanham, MD: Rowman & Littlefield.

Fraisse, P. (1982). The adaptation of the child to time. In W. J. Friedman (Ed.), *The developmental psychology of time* (pp. 113–140). New York, NY: Academic.

Friedman, K. C. (1944). Time concepts of elementary-school children. *The Elementary School Journal, 44*(6), 337–342.

Friedman, W. J. (1978). Development of time concepts in children. In H. W. Reese & L. P. Lipsitt (Eds.), *Advances in child development and behavior* (Vol. 12, pp. 267–298). New York, NY: Academic Press.

Friedman, W. J. (1982a). Introduction. In W. J. Friedman (Ed.), *The developmental psychology of time* (pp. 1–11). New York, NY: Academic Press.

Friedman, W. J. (1982b). Conventional time concepts and children's structuring of time. In W. J. Friedman (Ed.), *The developmental psychology of time* (pp. 171–208). New York, NY: Academic Press.

Friedman, W. J. (1990). *About time: Inventing the fourth dimension.* Cambridge (MA) / London: MIT Press.

Friedman, W. J. (1991). Children's memory for the time of past events. *Child Development, 62*(1), 139–155.

Friedman, W. J. (1992a). Children's time memory: The development of a differentiated past. *Cognitive Development* 7, 171–187.

Friedman, W. J. (1992b). Time memory and time perception. In F. Macar & V. Pouthas (Eds.), *Time, action and cognition: Towards bridging the gap.* Dordrecht, The Netherlands: Kluwer.

Friedman, W. J. (1993). Memory for the time of past events. *Psychological Bulletin, 113*(1), 44–66.

Friedman, W. J. (2005). Developmental and cognitive perspectives on humans' sense of the times of past and future events. *Learning and Motivation,* 36, 145–158.

Friedman, W. J. (2008). Developmental perspectives on the psychology of time. In S. Grondin (Ed.), *Psychology of time* (pp. 345–366). Bingley, UK: Emerald.

Frisby, D. (1992). *The alienated mind: The sociology of knowledge in Germany 1918–1933* (2nd ed.). London, UK: Routledge.

Frijhoff, W. (2010). The relevance of Dutch history, or: Much in little? Reflections of the practice of history in the Netherlands. *BMGN Low Countries Historical Review, 125*(2-3), 7–44.

Gadamer, H.-G. (1987). The problem of historical consciousness. In P. Rabinow & W.M. Sullivan (Eds.), *Interpretive social science: A second look* (pp. 82–140). Berkeley, CA: University of California Press.

Gautschi, P. (2009). *Guter Geschichtsunterricht* [Good history teaching]. Schwalbach im Taunus, Germany: Wochenschau Verlag.

Geary, P. J. (2002). *The myth of nations: The medieval origins of Europe.* Princeton, NJ: Princeton University Press.

Gieles, P. F. M. (1973). *Leerstofanalyse en het geschiedenisonderwijs* [Analysis of subject matter and history education]. 's-Hertogenbosch, the Netherlands: Katholiek Pedagogisch Centrum.

Gieles, P. H. F., & Greep, J. C. M. (1979). *Geschiedenis in de lagere klassen van het algemeen voortgezet onderwijs* [History in the first years of general secondary education]. The Hague, Netherlands: Verening van Geschiedenisleraren in Nederland (VGN).

Giesen, B. (2004). Noncontemporaneity, asynchronicity and divided memories. *Time & Society, 13*(1), 27–40.

Glöckel, H. (1973). *Geschichtsunterricht* [History Teaching]. Bad Heilbrunn, Germany: Verlag Julius Klinkhardt.

Greene, S. (1994). The problems of learning to think like a historian: Writing history in the culture of the classroom. *Educational Psychologist, 29*(2), 89–96.

Grever, M. (2007a). De natiestaat als pedagogische onderneming [The nation state as a pedagogical enterprise]. In M. Grever & K. Ribbens (Eds.), *Nationale identiteit en meervoudig verleden* (pp. 35–60). Amsterdam, The Netherlands: WRR/Amsterdam University Press.

Grever, M. (2007b). Veranderde coördinaten van herinnering' [Changing coordinates of memory]. In M. Grever & K. Ribbens (Eds.), *Nationale identiteit en meervoudig verleden* (pp. 61–86). Amsterdam, The Netherlands: WRR/ Amsterdam University Press.

Grever, M. (2009). Geen identiteit zonder oriëntatie in de tijd: Over de noodzaak van chronologie [No identity without orientation in time: About the necessity of chronology]. *BMGN Low Countries Historical Review, 124*(3), 438–451.

Grever, M., Jonker, E., Ribbens, K., & Stuurman, S. (2006). *Controverses rond de canon* [Controversies around the canon]. Assen, The Netherlands: Van Gorcum.

Grever, M., & Stuurman, S. (Eds.). (2007). *Beyond the canon: History for the twenty-first century.* New York, NY: Palgrave MacMillan.

Grondin, S. (2008). Introduction: A tentative picture of the ubiquity of time. In S. Grondin (Ed.), *Psychology of time* (pp. xi–xv). Bingley, UK: Emerald.

Grütter, H. T. (1994). Warum fasziniert die Vergangenheit? Perspektiven einer neuen Geschichtskultur' [Why does the past fascinate? Perspectives of a new historical culture]. In K. Füssmann, H.T. Grütter, & J. Rüsen (Eds.), *Historische faszination: Geschichtskultur heute* (pp. 45–57). Köln, Germany: Böhlau.

Günther-Arndt, H., & Sauer, M. (Eds.). (2006). *Geschichtsdidaktik empirisch: Untersuchungen zum historischen Denken und Lernen* [Teaching and learning history empirically explored: Researches in historical thinking and learning]. Berlin, Germany: Lit Verlag.

Hachtmann, R., Rohlfes, J., & Ullrich, V. (2006). *Deutsche geschichte. Wie wir wurden, was wir sind, 19. jahrhundert* [German history: How we became what we are, 19th century]. Stuttgart, Germany: Klett Verlag.

Hall, E. T. (1983). *The dance of life: The other dimension of time.* New York, NY: Anchor Books.

Hallam, R. N. (1970). Piaget and thinking in history. In M. Ballard (Ed.), *New movements in the study and teaching of history*. London, UK: Maurice Temple Smith.

Hallowell, A. I. (1937). Temporal orientation in western civilization and in a pre-literate society. *American Anthropologist*, New Series, *39*(4), 647–670.

Handelingen (2009), *Handelingen van de Tweede Kamer der Staten-Generaal* [Parliamentary Records of the Second Chamber of the Estates-General], 31972, no. 5, June 30, 2009.

Happold, F. C. (1927), *The study of history as a training in the art of thought*. London: Historical Association (Leaflet 69).

Happold, F. C. (1928). *The approach to history*. London: Christophers.

Harner, L. (1982). Talking about the past and the future. In W. J. Friedman (Ed.), *The developmental psychology of time* (pp. 141–169). New York, NY: Academic Press.

Harrison, M. L. (1934). The nature and development of concepts of time among young children. *The Elementary School Journal, 34*(7), 507–514.

Hartmann, U., & Hasselhorn, M. (2008). Historical perspective taking: A standardized measure for an aspect of students' historical thinking. *Learning and Individual Differences, 18*, 264–270.

Hartog, J. (1980). *Aruba, zoals het was, zoals het werd* [Aruba, how as it was and how it became] (2nd ed.). Aruba: Van Dorp.

Hartog, J. (1993). *De geschiedenis van twee landen. De Nederlandse Antillen en Aruba* [The history of two countries: the Netherlands Antilles and Aruba]. Zaltbommel, The Netherlands: Europese Bibliotheek.

HEG (werkgroep Herziening Eindexamen Geschiedenis) (1986). Werkgroep herziening eindexamen geschiedenis, *Deeladvies werkgroep HEG mavo/havo/vwo* [Working Group on Revision of the Final Exams for History, Advice on lower, intermediate and advanced secondary education]. The Hague, Netherlands: Vereniging van geschiedenisleraren in Nederland (VGN).

Heidegger, M. (1993). *Sein und Zeit*, 17. unveränderter Nachdruck des 15. [Being and Time, 17th unchanged printing of the 15th edition] Tübingen, Germany: Niemeyer. (Original work published 1927)

Hindsight, GCSE modern history review. Vol. 1–8 (1990–1998). Stratton, UK: Philip Allen Magazines.

Hodkinson, A. (2004). Does the English curriculum for history and its schemes of work effectively promote primary-aged children's assimilation of the concepts of historical time? Some observations based on current research. *Educational Research, 46*(2), 99–117.

Hoodless, P. A. (2002). An investigation into children's developing awareness of time and chronology in story. *Journal of Curriculum Studies, 34*(2), 173–200.

Howe, L. E. A. (1981). The social determination of knowledge: Maurice Bloch and Balinese time. *Man*, New Series, *16*(2), 220–234.

Howson, J. (2007). "Is it the Tuarts and then the Studors or the other way round?" The importance of developing a usable big picture of the past. *Teaching History, 127*, 40–47.

Husserl, E. G. A. (1928). *Vorlesungen zur phänomenologie des inneren zeitbewusstseins*, herausgegeben von Martin Heidegger [Lectures on the phenomenology of inner time consciousness, edited by Martin Heidegger]. Halle, Germany: Max Niemeyer Verlag.

Huizinga, J. (1948a). Uit de voorgeschiedenis van ons nationaal besef [From the prehistory of our national consciousness]. In *Verzamelde Werken* (vol. 2, pp. 97–160). Haarlem, The Netherlands: Tjeenk Willink.

Huizinga, J. (1948b). L'état bourguignon, ses rapports avec la France, et les origines d'une nationalité néerlandaise [The Burgundian state, its relations with France, and the origins of a Dutch nationality]. In *Verzamelde Werken* (vol. 2, pp. 161–215). Haarlem, The Netherlands: Tjeenk Willink.

Huizinga, J. (1950). De taak der cultuurgeschiedenis [The mission of cultural history]. In *Verzamelde Werken* (vol. 7, pp. 35–94). Haarlem, The Netherlands: Tjeenk Willink.

Jacott, L., López-Manjón, A., & Carretero, M. (1998). Generating explanations in history. In J. F. Voss & M. Carretero (Eds.), *Learning and reasoning in history: International review of history education* (vol. 2, pp. 294–306). London, UK: Woburn.

Jahoda, G. (1963). Children's concepts of time and history. *Educational Review, 15*(2), 87–104.

Jeissmann, K. -E. (1978). Friedrich Kohlrausch (1780–1867). In S. Quandt (Ed.), *Deutsche Geschichtsdidaktiker des 19. und 20. Jahrhunderts* (pp. 41–83). Paderborn, Germany: Schöningh.

Jeissmann, K. -E. (2000). Zum verhältnis von fachwissenschaft und fachdidaktik. [About the relation between academic disciplines and domain-specific educational science]. In K. –E. Jeissmann (Ed.), *Geschichte und bildung: Beiträge zur geschichtsdidaktik und zur historischen bildungsforschung* (pp. 73–86). Paderborn, Germany: Schöningh.

Jenkins, K. (1991). *Re-thinking history*. London, UK: Routledge.

Jones, G. E. (2000). The debate over the national curriculum for history in England and Wales 1989–1990: The role of the press. *The Curriculum Journal, 11*(3), 299–322.

Jonker, E. (2001a). *Historie: Over de blijvende behoefte aan geschiedenis* [History: About the Continuous Demand for Historical Insights]. Assen, The Netherlands: Van Gorcum.

Jonker, E. (2001b). Goede tijden, slechte tijden [Good times, bad times] In M. Grever & H. Jansen (Eds.), *De ongrijpbare tijd* (pp. 165–174). Hilversum, The Netherlands: Verloren.

Jonker, E. (2006). Sotto voce: Identiteit, burgerschap en de nationale canon [Sotto voce: Identity, citizenship and national canon]. *Tijdschrift voor Geschiedenis, 119*, 178–195.

Jordanova, L. (2006). *History in practice* (2nd ed.). London, UK: Hodder.

Kant, I. (1787). *Kritik der reinen Vernunft*, Zweite Auflage, [Critique of Pure Reason, 2nd ed.] Book I, part 1, section II, paragraph 4:1-2. Retrieved from http://gutenberg.spiegel.de/?id=5&xid=1369&kapitel=1#gb_found

Keatinge, M. W. (1910). *Studies in the teaching of history*. (s.l.): Black.

Kelty, M. G. (1925). Time expressions comprehended by children of elementary schools (part I, part II). *The Elementary School Journal 25*(7), 522–528, and *25*(8), 607–618.

Kocka, J. (1977). Gesellschaftliche funktionen der geschichtswissenschaft' [Societal functions of historical science]. In W. Oelmüller (Ed.), *Wozu noch geschichte?* (pp. 11–33). München, Germany: Wilhelm Fink.

Körber, A. (1998). Can our pupils stand in the shoes of someone else? In J. van der Leeuw-Roord (Ed.), *The state of history education in Europe* (pp. 123–138). Hamburg, Germany: Körber-Stiftung.

Körber, A. (2006). Hinreichende transparenz der operationen und modi des historische denkens im unterricht? [Sufficient transparancy in the operations and modes of historical thinking in education?]. In H. Günther-Arndt & M. Sauer (Eds.), *Geschichtsdidaktik empirisch: Untersuchungen zum historischen denken und lernen* (pp. 189–214). Münster, Germany: Lit Verlag.

Koselleck, R. (2004). *Futures past: On the semantics of historical time*. New York, NY: Columbia University Press.

LaCapra, D. (1998). *History and memory after Auschwitz*. Ithaca, NY: Cornell University Press.

Langer-Plän, M. (2003). Problem der quellenarbeit: Werkstattbericht aus einem empirischen projekt [The problem of source work: Report from a workshop in a project in empirical exploration]. *Geschichte in Wissenchaft und Unterricht, 54*, 319–336.

Lee, P. J. (1983). History teaching and philosophy of history. *History and Theory, 22*(4), 19–49.

Lee, P. (2004). "Walking backwards into tomorrow." Historical consciousness and understanding history. *International Journal of Historical Learning, Teaching and Research, 4*(1). Retrieved from http://www.heirnet.org/IJHLTR/journal7/lee.pdf

Lee, P. J. (2005). Putting principles into practice: Understanding history. In M. S. Donovan & J. D. Bransford (Eds.), *How students learn history in the classroom* (pp. 29–78). Washington, DC: National Academies Press.

Lee, P., & Ashby, R. (2000). Progression in historical understanding among students ages 7 – 14. In P. N. Stearns, P. Seixas, & S. Wineburg (Eds.), *Knowing, teaching, and learning history: National and international perspectives* (pp. 199–222). New York, NY: New York University Press.

Lee, P., & Ashby, R. (2001). Empathy, perspective taking and rational understanding. In O. L. Davis Jr., E.A. Yeager, & S.J. Foster (Eds.), *Historical empathy and perspective taking in the social studies* (pp. 21–50). Lanham, MD: Rowman & Littefield.

Lee, P., Ashby, R., & Dickinson, A. (1993, September). *Progression in children's ideas about history: Project CHATA (Concepts of History and Teaching Approaches: 7 to 14).* Paper presented at the Annual Conference of the British Educational Research Association, Liverpool, UK. Retrieved from http://eric.ed.gov/ERICDocs/data/ericdocs2sql/content_storage_01/0000019b/80/14/3b/39.pdf

Lee, P., Dickinson, A., & Ashby, R. (1997). "Just another emperor": Understanding action in the past. *International Journal of Educational Research, 27,* 233–244.

Lee, P., & Howson, J. (2009). "Two out of five did not know that Henry VIII had six wives: History education, historical literacy and historical consciousness. In L. Symcox & A. Wilschut (Eds.), *National history standards: The problem of the canon and the future of teaching history* (pp. 211–261). Charlotte, NC: Information Age Publishers.

Le Poidevin, R. (2003). *Travels in four dimensions.* Oxford, UK: Oxford University Press.

Lévesque, S. (2008). *Thinking historically: Educating students for the twenty-first century.* Toronto, ON: University of Toronto Press.

Levstik, L. (2000). Articulating the silences: Teachers and adolescents' conceptions of historical significance. In P. N. Stearns, P. Seixas, & S. Wineburg (Eds.), *Knowing, teaching and learning history: National and International perspectives* (pp. 284–305). New York, NY: New York University Press.

Levstik, L. S., & Barton, K. C. (2008). They still use some of their past: Historical salience in elementary children's chronological thinking. In L. S. Levstik & K. C. Barton (Eds.), *Researching history education* (pp. 108–147). New York, NY: Routledge.

Limón, M. (2002). Conceptual change in history. In M. Limón & L. Mason (Eds.), *Reconsidering conceptual change: Issues in theory and practice* (pp. 259–289). Dordrecht, Netherlands: Kluwer Academic Press.

Little, V. (1990). A national curriculum in history: A very contentious issue. *British Journal of Educational Studies, 38*(4), 319–334.

Loewe, M. (1999). Cyclische en lineaire concepten van tijd in China [Cyclic and linear concepts of time in China]. In K. Lippincott (Ed.), *Het verhaal van de tijd* (pp. 76–79). Haarlem, The Netherlands: Schuyt & Co. [English edition: *The story of time.* London, UK: Holberton 1999. The reference is to the Dutch edition.]

Lorenz, C. (1987). *De constructie van het verleden. Een inleiding in de theorie van de geschiedenis* [The construction of the past: An introduction to the theory of history]. Amsterdam, The Netherlands: BoomMeppel.

Lowenthal, D. (1985). *The past is a foreign country.* Cambridge, UK: Cambridge University Press.

Lowenthal, D. (1998). *The heritage crusade and the spoils of history.* Cambridge, UK: Cambridge University Press.

Lowyck, J., & Verloop, N. (1995). *Onderwijskunde: Een kennisbasis voor professionals* [Educational science: A knowledge base for professionals]. Groningen, The Netherlands: Wolters-Noordhoff.

Mannheim, K. (1952). The problem of generations. In K. Mannheim (Ed.), *Essays on the Sociology of Knowledge* (pp. 276–322). London, UK: Routledge & Kegan Paul.

Marsden, W. (1989). "All in a good cause": Geography, history and the politicization of the curriculum in 19th and 20th century England. *Journal of Curriculum Studies, 21*(6), 509–526.

Mayer, U., & Pandel, H. -J. (1976). *Kategorien der geschichtsdidaktik und praxis der unterrichtsanalyse* [Categories in expertise on teaching and learning history and the practice of educational analysis] Stuttgart, Germany: Ernst Klett Verlag.

McAulay, J. D. (1961). What understandings do second grade children have of time relationships? *Journal of Educational Research, 54*(8), 312–314.

McCormack, T., & Hoerl, C. (2008). Temporal decentering and the development of temporal concepts. *Language Learning, 58*(suppl. 1), 89–113.

McKeown, M. G., & Beck, I. L. (1990). The assessment and characterization of young learners' knowledge of a topic in history. *American Educational Research Journal, 27*(4), 688–726.

McTaggart, J. M. E. (1908). The unreality of time. *Mind, 17*, 457–474.

Mulder, E. (1990). Lessen in het vaderland: Geschiedenisonderwijs op de lagere school tot in de 20[e] eeuw [Lessons on the fatherland: History teaching in primary schools until the twentieth century]. *Holland, regionaal historisch tijdschrift, 22*, 216–224.

Nash, G. B., Crabtree, C., & Dunn, R. E. (1997). *History on trial: Culture wars and the teaching of the past.* New York, NY: Alfred A. Knopf.

Netzel, B. (n.d.). Zur geschichte der Henning von Burgsdorff-Stiftung [About the history of the Henning von Burgsdorff Foundation]. Retrieved from http://www.vgd-nds.de

Newton, I. (1687). *Principia mathematica philosophiae naturalis.* Via Gutenberg. Retrieved from http://www.gutenberg.org/ebooks/28233

Niemeijer, J. A. (2000). *J.H. Isings, historieschilder en illustrator* [J.H. Isings, painter of history and illustrator]. Kampen, The Netherlands: Kok.

Nietzsche, F. (1983). *Over nut en nadeel van geschiedenis voor het leven: Tweede traktaat tegen de keer* [On the utility and the liability of history for life: Second unfashionable observation]. Groningen, The Netherlands: Historische Uitgeverij (Original work published 1874)

Nooteboom, C. (2010, December 13). Europa en het werk der herinnering [Europe and the work of memory]. *NRC-Handelsblad*, p. 8.

Nuñez, R. E., & Sweetser, E. (2006). With the future behind them: Convergent evidence from Aymara language and gesture in the crosslinguistic comparison of spatial construals of time. *Cognitive Science, 30*, 401–450.

Oakden, E. C., & Sturt, M. (1922). The development of the knowledge of time in children. *British Journal of Psychology, 12*, 309–336.

Oakeshott, M. (1983). *On history and other essays.* Oxford, UK: Blackwell.

Ohnuki-Tierney, E. (1973). Sakhalin Ainu time reckoning. *Man,* New Series, *8*(2), 285–299.

Onderwijsraad. (2005). *De stand van educatief Nederland* [The state of education in the Netherlands]. The Hague, The Netherlands: Author.

Parker, R. A. (1950). *The calendars of ancient Egypt.* Chicago, IL: Oriental Institute of the University of Chicago.

Perfetti, C. A., Britt, M. A., & Georgi, M. C. (1995). *Text-based learning and reasoning: Studies in history.* Hillsdale, NJ: Lawrence Erlbaum.

Perkins, M. (1998). Timeless cultures: The "dreamtime" as colonial discourse. *Time and Society, 7,* 335–351.

Phillips, R. (1998). *History teaching, nationhood and the state.* London, UK: Cassell.

Piaget, J. (1969). *The child's conception of time.* London, UK: Routledge & Kegan Paul.

Picard, L. (1927). Moderne heldenverering: Napoleon en Bismarck [The modern cult of heroes: Napoleon and Bismarck]. *Den Gulden Winckel,* 26, 236–238.

Pinder, W. (1961). *Das Problem der Generationen in der Kunstgeschichte Europas* [The Problem of Generations in European History of Art], 2nd ed. Munich, Germany: Bruckmann.

Plato. (n.d.). *Timaeus,* Book 1. B. Jowett (Trans.). Retrieved from http://www.gutenberg.org/files/1572/1572-h/1572-h.htm#2H_4_0010

Plumb, J. H. (1969). *The death of the past.* London, UK: MacMillan.

Pocock, J. G. A. (1962). The origins of the study of the past: A comparative approach. *Comparative Studies in Society and History, 4*(2), 209–246.

Popp, S. (2006). Geschichtliches Überblickswissen aufbauen: Ein konzentrisch-longitudinales geschichtscurriculum aus den Niederlanden [Constructing historical survey knowledge: A concentric-longitudinal history curriculum from the Netherlands]. In E. Erdmann, R. Maier, & S. Popp (Eds.), *History teaching international* (pp. 269–301). Hanover, Germany: Georg Eckert Institut.

Popp, S. (2009). Orientierungswissen und nachhaltiges Lernen [Orientational knowledge and retentive learning]. *Geschichte in Wissenschaft und Unterricht, 60*(11), 646–657.

Popper, K. R. (1961). *The poverty of historicism.* 3rd ed. London: Routledge & Kegan Paul.

Poster, J. B. (1973). The birth of the past: Children's perception of historical time. *The History Teacher, 6*(4), 587–598.

Prangsma, M. E. (2007). *Multimodal representations in collaborative history learning.* Unpublished dissertation. University of Utrecht, Utrecht, the Netherlands.

Price, M. (1968). History in danger. *History, 53,*(179), 342–347.

Raad van State (2009). *Advies inzake het Besluit houdende wijziging van het Besluit vernieuwde kerndoelen WPO en het Besluit kerndoelen onderbouw VO in verband met invoering van de canon in de kerndoelen.* [Council of Sate, Advice on the Decree concerning amending the renewed attainment targets for primary education and first stage secondary education in relation to the introduction of the canon into the attainment targets]. The Hague, Netherlands: Staatscourant 103, 9 June 2009.

Reeves, M. (1980). *Why history?* London, UK: Longman.

Ribbens, K. (2002). *Een eigentijds verleden: Alledaagse historische cultuur in Nederland 1945–2000* [A contemporary past: Quotidian historical culture in the Netherlands]. Hilversum, The Netherlands: Verloren.

Ricoeur, P. (1984). *Time and narrative* (Vol. 1). Chicago, IL: University of Chicago Press.

Ricoeur, P. (1988). *Time and narrative* (Vol. 3). Chicago, IL: University of Chicago Press.

Roeckelein, J. E. (2008). History of conceptions and accounts of time and early time perception research. In S. Grondin (Ed.), *Psychology of time* (pp. 1–50. Bingley, UK: Emerald.

Romein, J. (1976). Duur en verandering in de geschiedenis: Het probleem van de 'omslag' [Duration and change in history: About the problem of 'turnings']. In J. Romein (Ed.), *Historische lijnen en patronen* (pp. 586–599). Amsterdam, The Netherlands: Querido.

Royzman, E. B., Cassidy, K. W., & Baron, J. (2003). "I know, you dnow: Epistemic egocentrism in children and adults. *Review of General Psychology, 7*(1), 38–65.

Rüsen, J. (1975). Zum verhältnis von theorie und didaktik der geschichte [On the relation between theory and history education]. *Geschichte in Wissenschaft und Unterricht 26*(8), 436–455.

Rüsen, J. (1994). *Historisches lernen* [Historical learning]. Köln, Germany: Böhlau.

Russell, B. (2000). *Autobiography.* New York, NY: Routledge.

Schmid, H. -D. (1972). Entwurf einer geschichtsdidaktik der mittelstufe (Design of a history curriculum for the middle school]. In H. Süssmuth (Ed.), *Geschichtsdidaktik ohne Zukunft?* (Vol. 1, pp. 208–245). Stuttgart, Germany: Klett Verlag.

Schmid, H. -D. (Ed.). (1974). *Fragen an die geschichte* [Questions to history] (Vols. 1–4). Frankfurt, Germany: Hirschgraben Verlag.

Schools Council History 13–16 Project. (1976a). *A new look at history.* Edinburgh, Scotland: Holmes MacDougal.

Schools Council History 13–16 Project. (1976b). *What is history?* (vol. 1 – 5). Edinburgh, Scotland: Holmes MacDougal.

Schuthof, P. (1983). Inleiding. [Introduction]. In H. Bergman (Ed.), *Onderwijskundigen van de twintigste eeuw* [Educationalists from the Twentieth Century]. Amsterdam, The Netherlands: Intermediair/Wolters-Noordhoff.

Schütz, A. (1972). *The phenomenology of the social world.* London, UK: Heinemann Educational Books.

Seixas, P. (1994). When psychologists discuss historical thinking: A historian's perspective. *Educational Psychologist, 29*(2), 107–109.

Seixas, P. (1996). Conceptualizing the growth of historical understanding. In D. R. Olson & N. Torrance (Eds.), *The Handbook of Education and Human Development. New Models of Learning, Teaching and Schooling,* p. 765-783. Oxford / Cambridge (MA): Blackwell.

SEP (Stanford Encyclopedia of Philosophy). (2002, November 25). Time. (Substantive revision Feb 26, 2008). Retrieved from http://plato.stanford.edu/entries/time

Shadish, W. R., Cook, Th. D., & Campbell D. T. (2002). *Experimental and quasi-experimental designs for generalized causal inference.* Boston, MA: Houghton Mifflin.

Shemilt, D. (2000). The caliph's coin: The currency of narrative frameworks in history teaching. In P. N. Stearns, P. Seixas, & S. Wineburg (Eds.), *Knowing, teaching and learning history: National and international perspectives* (pp. 83–101). New York, NY: New York University Press.

Shulman, L. S. (2004). Those who understand: Knowledge growth in teaching. In L. S. Shulman (Ed.), *The wisdom of practice: Essays on teaching, learning, and learning to teach* (pp. 187–215). San Francisco, CA: Jossey-Bass.

Sniekers, J. (1993). *Een leerplan geschiedenis en staatsinrichting* [A curriculum for history and polity]. Groningen, The Netherlands: Wolters-Noordhoff.

SPTF (Stuurgroep Profiel Tweede Fase). (1995). *Advies examenprogramma's havo en vwo: Geschiedenis en staatsinrichting, aardrijkskunde, maatschappijleer* [Working group on social studies in upper secondary schools: Advice on exam programs for history, geography and social studies]. Enschede, The Netherlands: Stichting Leerplan Ontwikkeling.

Stephens, W. B. (1998). *Education in Britain 1750–1914.* London, UK: MacMillan Press.

Stow, W., & Haydn, T. (2000). Issues in the teaching of chronology. In J. Arthur & R. Phillips (Eds.), *Issues in history teaching* (pp. 83–97). London, UK: RoutledgeFalmer.

Süssmuth, H. (Ed.) (1972). *Geschichtsunterricht ohne zukunft?* [History education without future?] (2 vols.). Stuttgart, Germany: Klett Verlag.

Sylvester, D. (1994). Change and continuity in history teaching 1900–93. H. Bourdillon (Ed.), *Teaching history* (pp. 9–23). London, UK: Routledge.

Symcox, L. (2002). *Whose history? The struggle for national standards in American classrooms.* New York, NY: Teacher's College Press.

Symcox, L., & Wilschut, A. (Eds.). (2009). *National history standards: The problem of the canon and the future of teaching history.* Charlotte, NC: Information Age Publishing.

Taylor, H. A. (1993). About time, inventing the fourth dimension by W.J. Friedman. *Philosophical Psychology, 6*(3), 341–342.

Teitler, H. (1998). *De opstand der Batavieren* [The revolt oft he Batavians]. Hilversum, The Netherlands: Verloren.

Thornton, S. J., & Vukelich, R. (1988). Effects of children's understanding of time concepts on historical understanding. *Theory and Research in Social Education, 16*, 69–82.

Toebes, J. G. (1976). Van een leervak naar een denk- en doevak: Een bijdrage tot de geschiedenis van het Nederlands geschiedenisonderwijs [From learning subject to thinking and doing subject: A contribution to the history of Dutch history education], *Kleio* 17, 202-284.

Toebes, J. G. (1981). *Geschiedenis: Een vak apart? Het probleem van de verbinding van geschiedenis met andere mens- en maatschappijvakken in het voortgezet onderwijs van de Bondsrepubliek Duitsland, Engeland en Nederland* [History: A subject by itself? The problem of the connection of history to other social sciences in secondary education in Germany, England and the Netherlands]. Meppel, The Netherlands: Krips Repro.

Toebes, J. G. (1983). *Een kleine didactiek geschiedenis onderbouw: Achtergronden van 'Vragen aan de geschiedenis'* [A comprehensive methodology of history education in the foundation years of secondary education: Backgrounds to 'Questions to history']. Groningen, The Netherlands: Wolters-Noordhoff.

Toebes, J. G. (1989). *Basisvorming geschiedenis en staatsinrichting* [The foundation years in teaching history and polity]. Leiden, The Netherlands: Stenfert Kroese/Martinus Nijhoff.

Toebes, J.G., e.a. (1983–1996), *Vragen aan de geschiedenis* [Questions to History] (schoolbook series). Groningen, The Netherlands: Wolters-Noordhoff.

Totten, S., Bartrop, P. R., & Jacobs, S. L. (Eds.). (2004). *Teaching about the Holocaust: Essays by college and university teachers.* Westport, CT: Praeger.

Turetzky, Ph. (1998). *Time.* London, UK: Routledge.

Turton, D., & Ruggles, C. (1978). Agreeing to disagree: The measurement of duration in a southwestern Ethiopian community. *Current Anthropology, 19*(3), 585–600.

VGD (Verband der Geschichtslehrer Deutschlands). (2006). *Bildungsstandards geschichte: Rahmenmodell gymnasium 5–10. jahrgangsstufe* [Association of history teachers in Germany, educational standards for history: An outline model for the gymnasium, grades 5–10]. Schwalbach, Germany: Wochenschau.

VGN (Dutch Association of History Teachers). (1967). Rapport van de Doelstellingencommissie, ingesteld door het Bestuur der V.G.N. [Report of the Commission on Educational Objectives, ordained by the Board of the], *Kleio, 8*(6), 161–188.

Van Boxtel, C. (1999a). Kleio-enquête naar de inhoud van een canon van chronologische kennis [Survey towards the contents of a canon of historical knowledge]. *Kleio, 40*(1), 28–30.

Van Boxtel, C. (1999b). De resultaten van de enquête chronologische kennis: Wat, waarom en hoe?' [The results of the survey towards chronological knowledge: What, why and how?], *Kleio, 40*(4), 2–11.

Van Boxtel, C., & Van Drie, J. (2004). Historical reasoning: A comparison of how experts and novices contextualise historical sources. *International Journal of Historical Learning, Teaching and Research, 4*(2), 89–97.

Van der Dussen, W. J. (1988). *Filosofie van de geschiedwetenschappen* [Philosophy of historical sciences]. Leiden, The Netherlands: Martinus Nijhoff.

Van der Werff, J. H. (1973). Van aanvaarding naar keuze: Dertien jaar Kleio over doelstellingen van het geschiedenisonderwijs [From acceptation to choice: Thirteen years of Kleio on the objectives of history education]. *Kleio, 14*(10), 432–439.

Van Drie, J. (2005). *Learning about the past with new technologies: Fostering historical reasoning in computer-supported collaborative learning.* Unpublished dissertation. University of Utrecht, Utrecht, the Netherlands.

Van Drie, J. P., & Van Boxtel, C. A. M. (2008). Historical reasoning: Towards a framework for analyzing students' reasoning about the past. *Educational Psychology Review, 20*(2), 87–110.

Van Oostrom, F. (2006). *Entoen.nu, Rapport van de Commissie Ontwikkeling Nederlandse Canon* [Report of the Committee for the Development of the Netherlands Canon] (2 vols.). Amsterdam, The Netherlands: Amsterdam University Press.

Van Sas, N. (2009). Geschiedenis—Herinnering—Identiteit: De historici en het nationaal historisch museum [History, memory, identity: Historians and the national historical museum]. *BMGN Low Countries Historical Review, 124*(3), 419–430.

Van Vree, F. (2009). De mythe van Nederland [The myth of the Netherlands]. *Tijdschrift voor Geschiedenis, 122,* 384–387.

Vann, R. T. (1998). The reception of Hayden White. *History and Theory, 37*(2), 143–161.

VanSledright, B. A. (2001). From empathetic regard to self-understanding: Im/Positionality, empathy and historical contextualization. In O. L. Davis Jr., E. A. Yeager, & S. J. Foster (Eds.), *Historical empathy and perspective taking in the social studies* (pp. 51–68. Lanham, MD: Rowman & Littlefield.

Vermeulen, E. E. G. (1978). *Waarden en geschiedwetenschap: Een vergelijking van de standpunten ingenomen door H. W. von der Dunk, A. G. Weiler, M. C. Brands, met notities over die van J. M. Romein, G. Harmsen, J. H. J. van der Pot.* [Values and historical science: A comparison of the views adopted by H. W. von der Dunk, A. G. Weiler, M. C. Brands, with notes about those adopted by J. M. Romein, G. Harmsen, J. H. J. van der Pot.]. Assen, The Netherlands: Van Gorcum.

Von Borries, B. (1987). "Reifung" oder "sozialisation" des geschichtsbewusstseins? Zur rekonstruktion einer vorschnell verschütteten kontroverse [Maturation or socialization? About the reconstruction of a prematurely concluded controversy]. *Geschichtsdidaktik, 12*(2), 143–159.

Von Borries, B., Pandel, H.-J., & Rüsen, J. (Eds.). (1991). *Geschichtsbewusstsein empirisch* [Historical consciousness empirically explored]. Pfaffenweiler: Centaurus.

Von Ranke, L. (1885), *Geschichte der romanischen und germanischen Völker von 1494 bis 1514*, 3e Auflage. [History of the Germanic and Romance peoples from 1494 until 1514] (3rd ed.). Leipzig, Germany: Verlag Von Duncker & Humblot.

Vos, C. (2009). Op zoek naar een Nederlandse identiteit. De verbeelding van de vaderlandse geschiedenis in Verleden van Nederland en Nederland in twaalf moorden [In search of Dutch identity: The imaging of fatherland history in 'The Netherlands Past' and 'The Netherlands in Twelve Murders']. *Tijdschrift voor Geschiedenis,* 122, 378–383.

Vukelich, R., & Thornton, S. J. (1990). Children's understanding of historical time: Implications for instruction. *Childhood Education, 67*(1), 22–25.

Vries, P. H. H. (1990). *Vertellers op drift. Een verhandeling over de nieuwe verhalende geschiedenis* [Narrators adrift. Observations on the new narrative history]. Hilversum, The Netherlands: Verloren.

WIEG (Werkgroep Implementatie Eindexamen Geschiedenis). (1993). *Geschiedenis, wat is dat? Een introductie op het nieuwe eindexamen geschiedenis en staatsinrichting* (Working Group on Implementation of Final Exams for History: History, what is it? An introduction to the new final exams in history]. The Hague, The Netherlands: Nijgh en Van Ditmar.

Walsh, W. H. (1967). Colligatory concepts in history. In W. H. Burston & D. Thompson (Eds.), *Studies in the nature and teaching of history* (pp. 65–84). London, UK: Humanities Press.

Ward, S. (2004). The end of empire and the fate of Britishness. In H. Brocklehurst & R. Phillips (Eds.), *History, nationhood and the question of Britain* (pp. 242–258). New York, NY: Palgrave MacMillan.

White, H. (1973). *Metahistory: The historical imagination in nineteenth-century Europe.* Baltimore, MD: Johns Hopkins University Press.

White, H. (1985). The historical text as literary artifact. In H. White (Ed.), *Tropics of discourse: Essays in cultural criticism* (pp. 81–100). Baltimore, MD: Johns Hopkins University Press (Oorspr. 1978).

White, H. (1987). *The content of the form: Narrative discourse and historical representation.* Baltimore, MD: Johns Hopkins University Press.

Wiegrefe, K., & Pieper, D. (Eds.). (2007). *Die erfindung der Deutschen: Wie wir wurden, was wir sind* [The invention of the Germans: How we became what we are]. München, Germany: Deutsche Verlags-Anstalt.

Wiesemüller, G. (1972). *Unbewältigte vergangenheit: Überwältigende gegenwart?* [A past not mastered: A present overwhelming?]. Stuttgart, Germany: Klett Verlag.

Williams, C. D. (2009). *Boudicca and her stories: Narrative transformations of a warrior queen.* Newark, DE: University of Delaware Press.

Wils, K. (2009). The evaporated canon and the overvalued source. In L. Symcox & A. Wilschut (Eds.), *National history standards: The problem of the canon and the future of teaching history* (pp. 15–31). Charlotte, NC: Information Age Publishers.

Wilschut, A. (2005). Het Nederlands slavernijverleden en het canonieke denken over geschiedenisonderwijs [The Dutch slavery past and canonical thinking about history education]. *Kleio, 46*(8), 34–38.

Wilschut, A. (2009). Canonical standards or orientational frames of reference? The cultural and the educational approach to the debate about standards in history teaching. In L. Symcox & A. Wilschut (Eds.), *National history standards: The problem of the canon and the future of teaching history* (pp. 117–140). Charlotte, NC: Information Age.

Wilschut, A. (2010). History at the mercy of politicians and ideologies: Germany, England, and the Netherlands in the 19th and 20th centuries. *Journal of Curriculum Studies, 42*(5), 693–723.

Wilschut, A., van Straaten, D., & van Riessen, M. (2004). *Geschiedenisdidactiek* [Teaching history]. Bussum, The Netherlands: Coutinho.

Wineburg, S. (1991). Historical problem solving: A study of the cognitive processes used in the evaluation of documentary and pictorial evidence. *Journal of Educational Psychology, 83,* 73–87.

Wineburg, S. (1998). Reading Abraham Lincoln: An expert/expert study in the interpretation of historical texts. *Cognitive Science, 22,* 319–346.

Wineburg, S. (2001). *Historical thinking and other unnatural acts.* Philadelphia, PA: Temple University Press.

Zerubavel, E. (1977). The French republican calendar: A case study in the sociology of time. *American Sociological Review, 42,* 868-877.

Zerubavel, E. (1996). Lumping and splitting: Notes on social classification. *Sociological Forum, 11*(3), 421–433.

Zerubavel, E. (2003). *Time maps: Collective memory and the social shape of the past.* Chicago, IL: University of Chicago Press.

APPENDIX A

PowerPoint Presentations Used in the Lessons According to the Chronological and Imaginative-Associative Approaches

1. Chronological

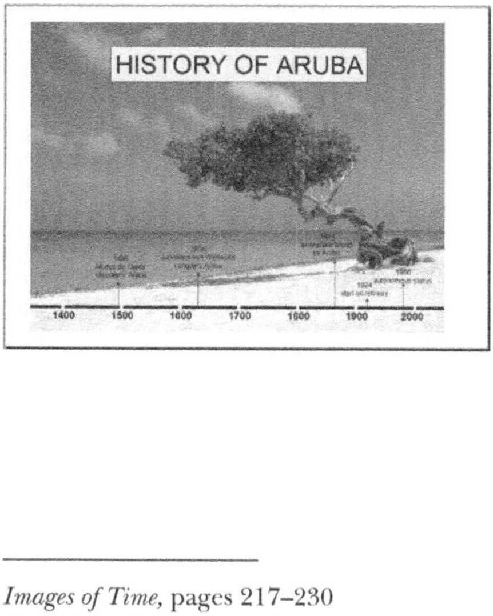

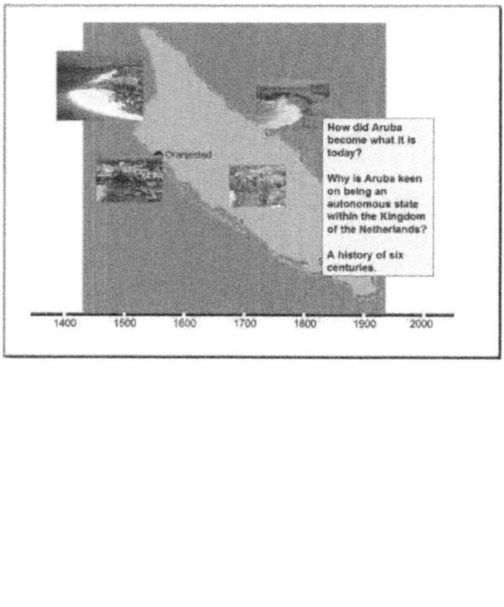

Images of Time, pages 217–230

1400
The
15th
century
1500

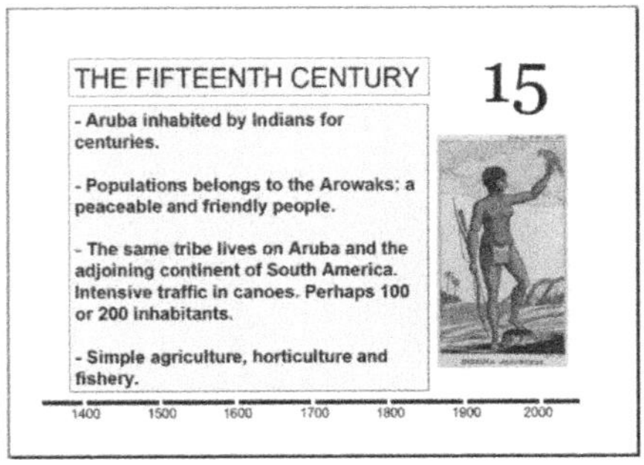
THE FIFTEENTH CENTURY
15
- Aruba inhabited by Indians for centuries.
- Populations belongs to the Arowaks: a peaceable and friendly people.
- The same tribe lives on Aruba and the adjoining continent of South America. Intensive traffic in canoes. Perhaps 100 or 200 inhabitants.
- Simple agriculture, horticulture and fishery.
1400 1500 1600 1700 1800 1900 2000

THE FIFTEENTH CENURY
15
Urn found on Aruba
Rock drawings found on op Aruba
1400 1500 2000

THE FIFTEENTH CENTURY
15
1499
The Spaniard
Alonso de Ojeda
'discovers' Aruba
1400 1500 1600 1700 1800 1900 2000

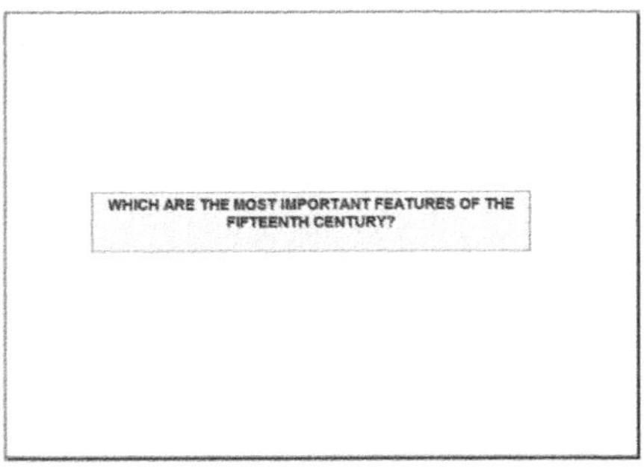
WHICH ARE THE MOST IMPORTANT FEATURES OF THE FIFTEENTH CENTURY?

1500
The
16th
century
1600

THE SIXTEENTH CENTURY
16
- 1513 Spaniards declare Aruba, Bonaire and Curaçao 'useless islands'.
- 1515 Aruba's complete population is deported.
- 1530 Spaniards import
goats. The island becomes
- Indians
appear to be
1400 1500

THE SIXTEENTH CENTURY
16
- The Spaniards convert the Indians to catholicism.
- In the middle of the island a wooden cross is put in place: 'Santa Cruz'.
1400 1500 1600 1700 1800 1900 2000

WHICH ARE THE MOST IMPORTANT FEATURES OF THE SIXTEENTH CENTURY?

1600 **The 17th century** 1700

THE SEVENTEENTH CENTURY 17

- ca 1600 English and Dutch arrive in the Carribean.
- 1634 Johan van Walbeeck conquers Curaçao.
- 1636 Aruba also occupied by Dutch with very few soldiers.
- Possession of the West India Company. Aruba provides horses and goats for Curaçao.

1400 1500 1600 1700 1800 1900 2000

THE SEVENTEENTH CENTURY 17

1400 1500 1600 1700 1800 1900 2000

WHICH ARE THE MOST IMPORTANT FEATURES OF THE SEVENTEENTH CENTURY?

1700 **The 18th century** 1800

THE EIGHTEENTH CENTURY 18

- Aruba remains possession of WIC.
- 1750 The Indians get their first simple church.
- 1754 The first white colonist (Moses Maduro) on Aruba.
- Almost no slaves on Aruba, in contrast with Curaçao: in 1795: thirty slaves on Aruba, almost thirteen thousand on Curaçao.

Alto Vista Chapel built in 1950 at the location of the first Indian church.

1400 1500 1600 1700 1800 1900 2000

THE EIGHTEENTH CENTURY 18

- Trade in 'Bay of Horses' increases.
- Frequent attacks by sea robbers and pirates.

- The Company calculates costs and gains: defence is expensive!
- The has about one thousand inhabitants, mainly Indians.

1400 1500 1600 1700 1800 1900 2000

THE EIGHTEENTH CENTURY
18
1796/'97 At last a fort is built: Fort Zoutman. It is the core of the later Oranjestad.
Fort Zoutman on Aruba. The tower has been added later, in the time of King William III (1848-1890). That's why it is called 'William III Tower'.
1400 1500 1600 1700 1800 1900 2000

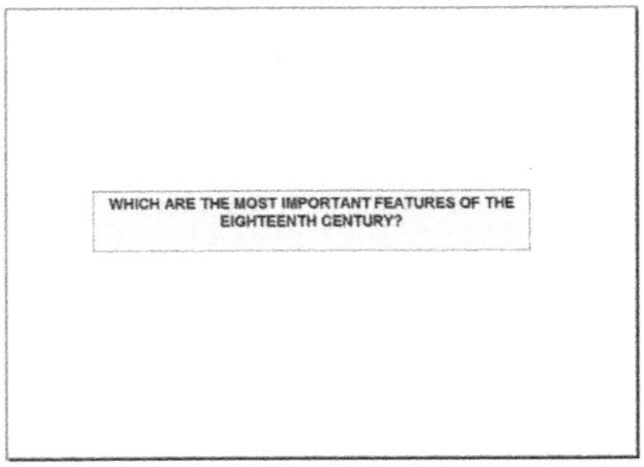
WHICH ARE THE MOST IMPORTANT FEATURES OF THE EIGHTEENTH CENTURY?

1800
The 19th century
1900

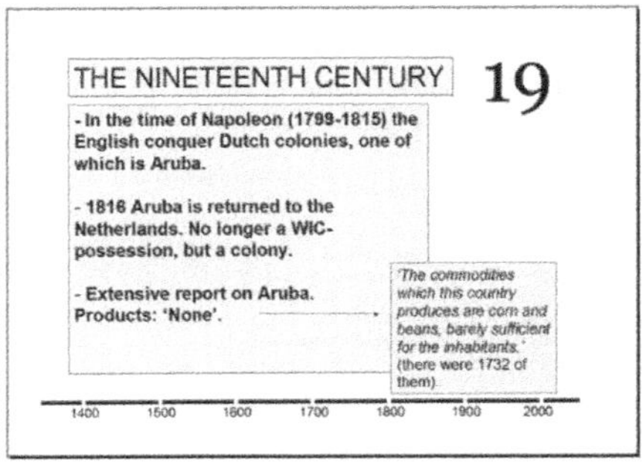
THE NINETEENTH CENTURY
19
- In the time of Napoleon (1799-1815) the English conquer Dutch colonies, one of which is Aruba.
- 1816 Aruba is returned to the Netherlands. No longer a WIC-possession, but a colony.
- Extensive report on Aruba. Products: 'None'.
'The commodities which this country produces are corn and beans, barely sufficient for the inhabitants.' (there were 1732 of them)
1400 1500 1600 1700 1800 1900 2000

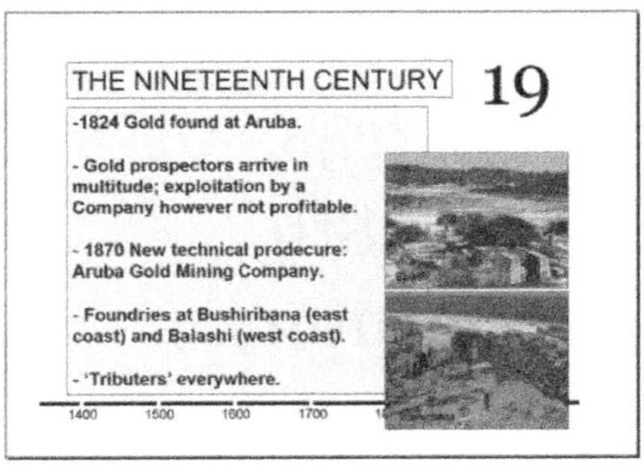
THE NINETEENTH CENTURY
19
-1824 Gold found at Aruba.
- Gold prospectors arrive in multitude; exploitation by a Company however not profitable.
- 1870 New technical prodecure: Aruba Gold Mining Company.
- Foundries at Bushiribana (east coast) and Balashi (west coast).
- 'Tributers' everywhere.
1400 1500 1600 1700

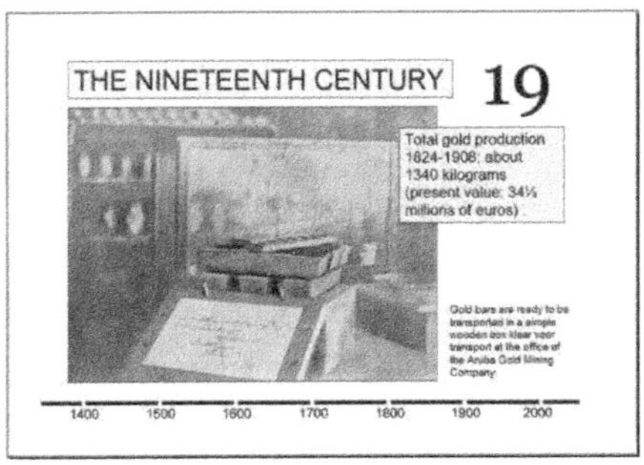
THE NINETEENTH CENTURY
19
Total gold production 1824-1908: about 1340 kilograms (present value: 34½ millions of euros)
1400 1500 1600 1700 1800 1900 2000

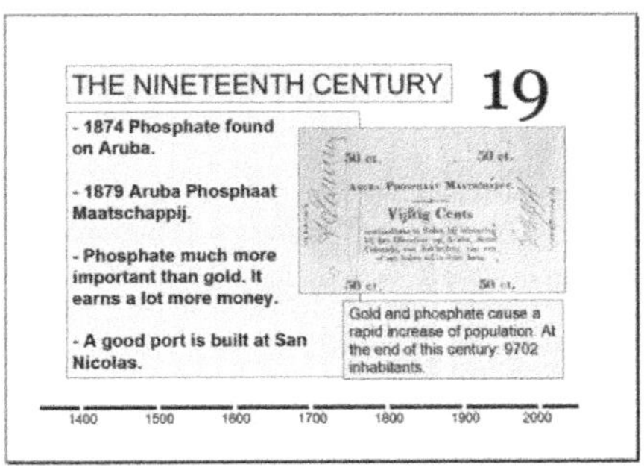
THE NINETEENTH CENTURY
19
- 1874 Phosphate found on Aruba.
- 1879 Aruba Phosphaat Maatschappij.
- Phosphate much more important than gold. It earns a lot more money.
- A good port is built at San Nicolas.
50 ct. 50 ct.
Vijftig Cents
50 ct. 50 ct.
Gold and phosphate cause a rapid increase of population. At the end of this century: 9702 inhabitants.
1400 1500 1600 1700 1800 1900 2000

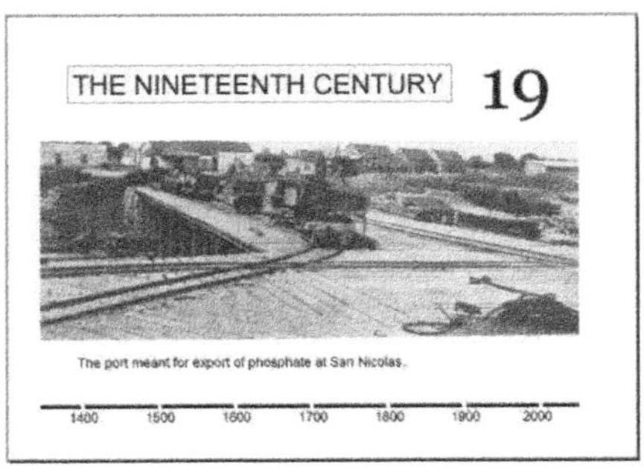
THE NINETEENTH CENTURY
19
The port meant for export of phosphate at San Nicolas.
1400 1500 1600 1700 1800 1900 2000

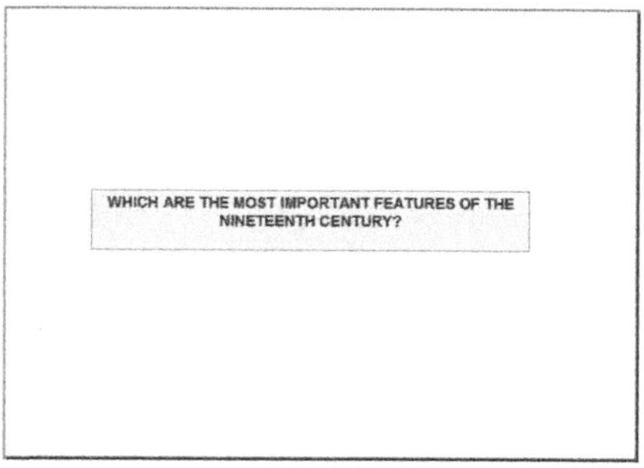
WHICH ARE THE MOST IMPORTANT FEATURES OF THE NINETEENTH CENTURY?

1900
The 20th century
2000

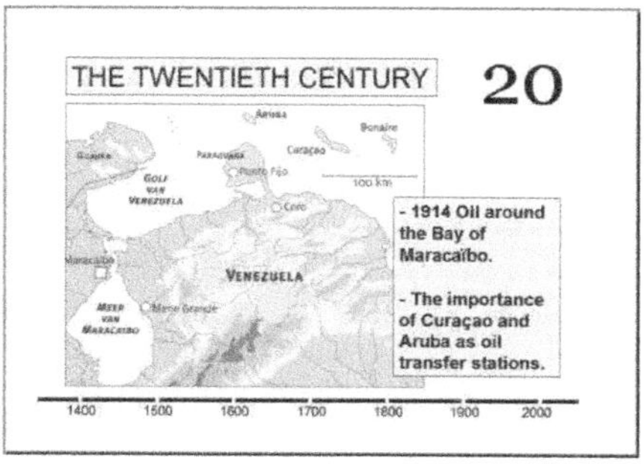
THE TWENTIETH CENTURY
20
VENEZUELA
- 1914 Oil around the Bay of Maracaïbo.
- The importance of Curaçao and Aruba as oil transfer stations.
1400
1500
1600
1700
1800
1900
2000

THE TWENTIETH CENTURY
20
- 1924 The visit by Captain Robert Rodgers.
- Governor Quast agrees to a tanker as a transfer station moored at San Nicolas.
- 1927 Port of San Nicolas fit to accomodate large oil tankers.
1800

THE TWENTIETH CENTURY
20
In 1927 it was decided that a refinery would be built.
1400
1500
1600
1700
1800
1900
2000

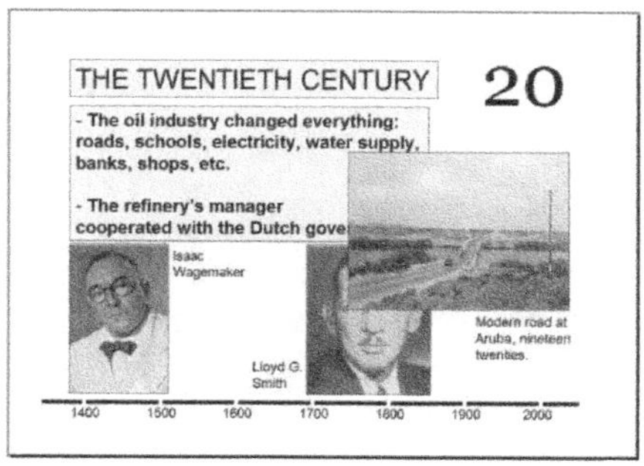
THE TWENTIETH CENTURY
20
- The oil industry changed everything: roads, schools, electricity, water supply, banks, shops, etc.
- The refinery's manager cooperated with the Dutch gove
Isaac Wagemaker
Lloyd G. Smith
Modern road at Aruba, nineteen twenties.
1400
1500
1600
1700
1800
1900
2000

THE TWENTIETH CENTURY
20
- Strategic importance of refinery in World War II: 6% of all airplane fuel was produced here.
- 1942 Attacked
Torpedoed tanker
1800
1900
2000

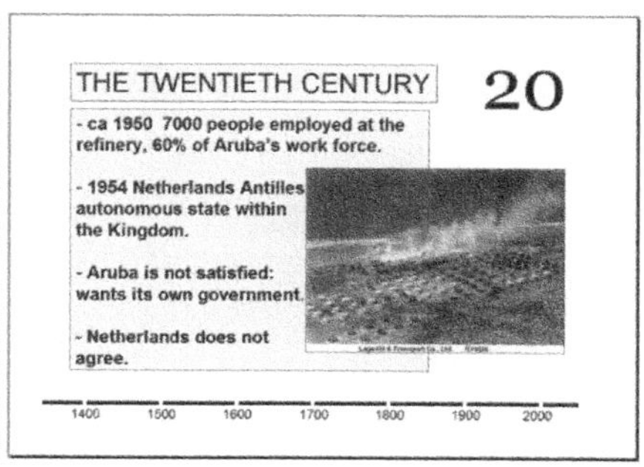
THE TWENTIETH CENTURY
20
- ca 1950 7000 people employed at the refinery, 60% of Aruba's work force.
- 1954 Netherlands Antilles autonomous state within the Kingdom.
- Aruba is not satisfied: wants its own government.
- Netherlands does not agree.
1400
1500
1600
1700
1800
1900
2000

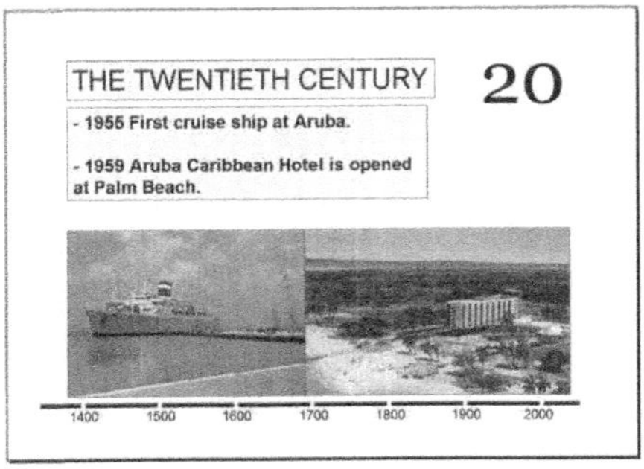
THE TWENTIETH CENTURY
20
- 1955 First cruise ship at Aruba.
- 1959 Aruba Caribbean Hotel is opened at Palm Beach.
1400 1500 1600 1700 1800 1900 2000

THE TWENTIETH CENTURY
20
- 1971 Betico Croes founds Movimiento Electoral di Pueblo (MEP, People's Electoral Movement) and strives for what he calls 'separate status'.
1400 1500 1600 1700 1800 1900 2000

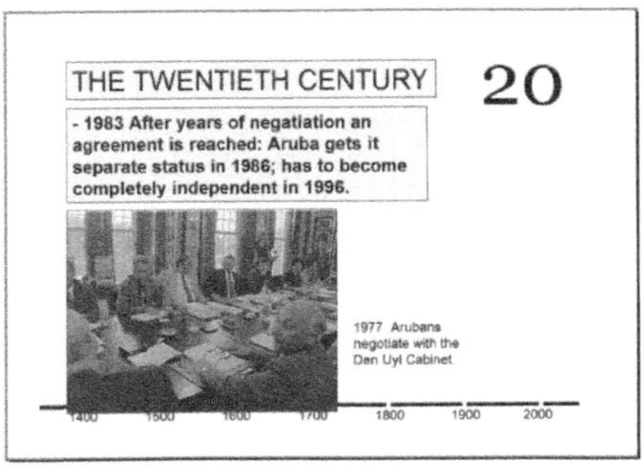
THE TWENTIETH CENTURY
20
- 1983 After years of negatiation an agreement is reached: Aruba gets it separate status in 1986; has to become completely independent in 1996.
1977 Arubans negotiate with the Den Uyl Cabinet
1800 1900 2000

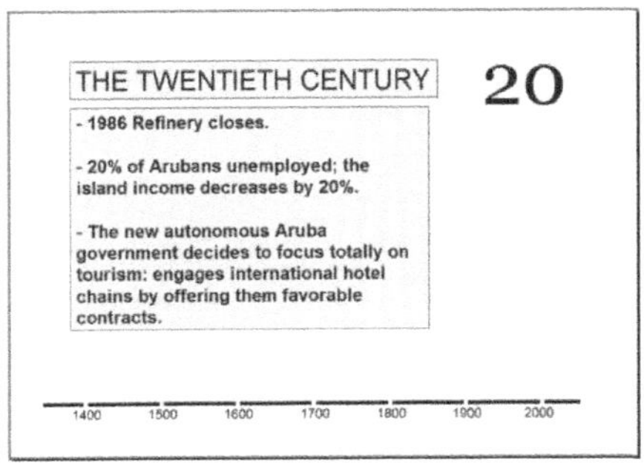
THE TWENTIETH CENTURY
20
- 1986 Refinery closes.
- 20% of Arubans unemployed; the island income decreases by 20%.
- The new autonomous Aruba government decides to focus totally on tourism: engages international hotel chains by offering them favorable contracts.
1400 1500 1600 1700 1800 1900 2000

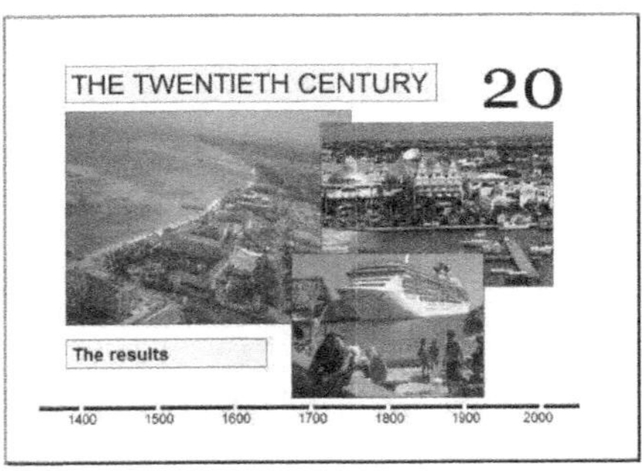
THE TWENTIETH CENTURY
20
The results
1400 1500 1600 1700 1800 1900 2000

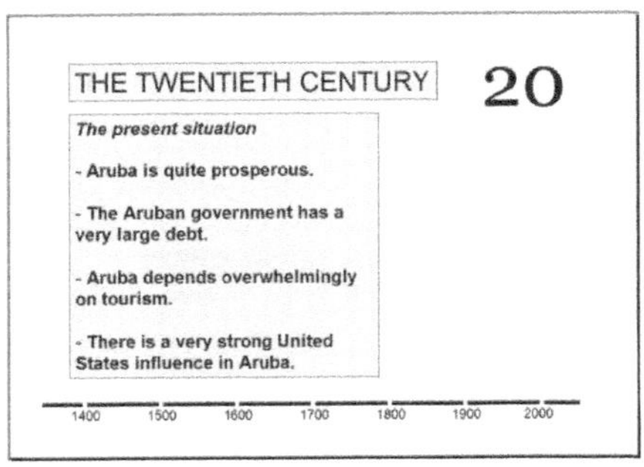
THE TWENTIETH CENTURY
20
The present situation
- Aruba is quite prosperous.
- The Aruban government has a very large debt.
- Aruba depends overwhelmingly on tourism.
- There is a very strong United States influence in Aruba.
1400 1500 1600 1700 1800 1900 2000

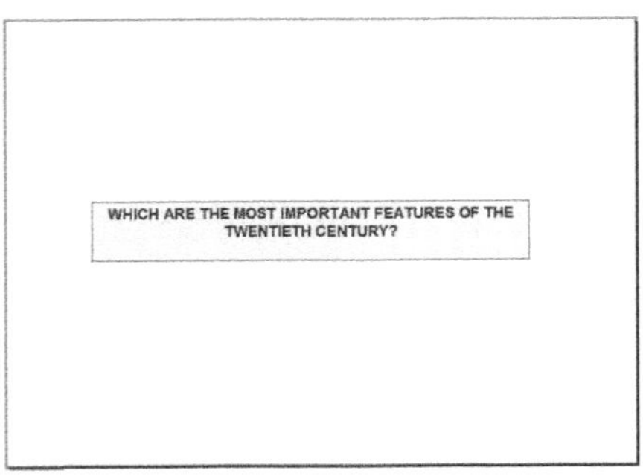
WHICH ARE THE MOST IMPORTANT FEATURES OF THE TWENTIETH CENTURY?

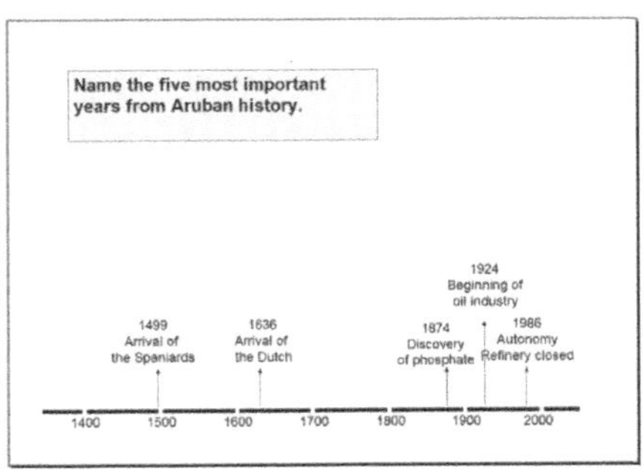
Name the five most important years from Aruban history.
1499 Arrival of the Spaniards
1636 Arrival of the Dutch
1874 Discovery of phosphate
1924 Beginning of oil industry
1986 Autonomy Refinery closed
1400 1500 1600 1700 1800 1900 2000

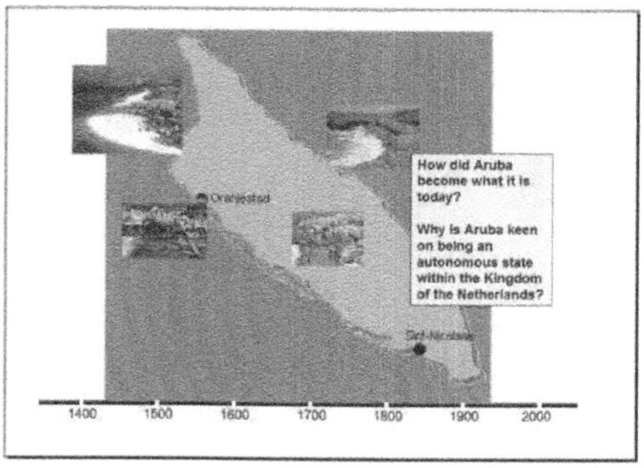

2. Imaginative-Associative

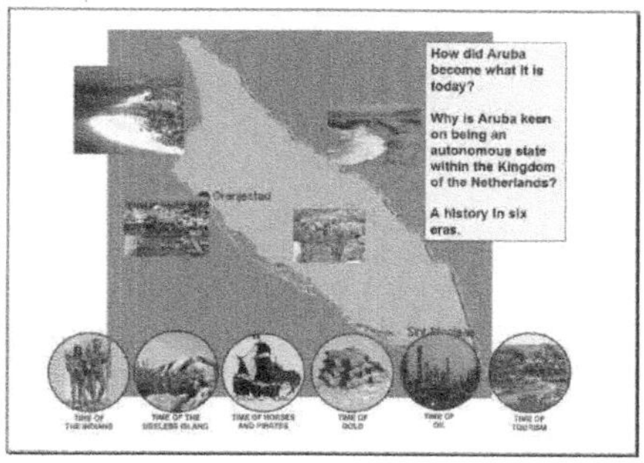

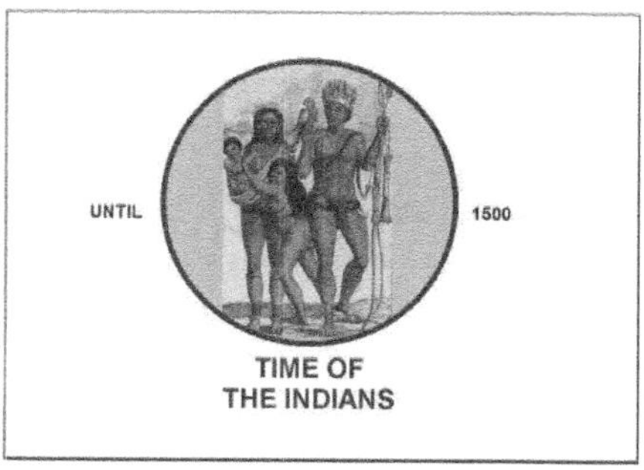

Excavation at
Tanki Flip,
Aruba.

Rock
Arowak
Indians.

WHAT COMES TO YOUR MIND ABOUT THE TIME OF THE INDIANS?

1500
1600
TIME OF THE
USELESS ISLAND

The whole population of Aruba
is deported by the Spaniards.

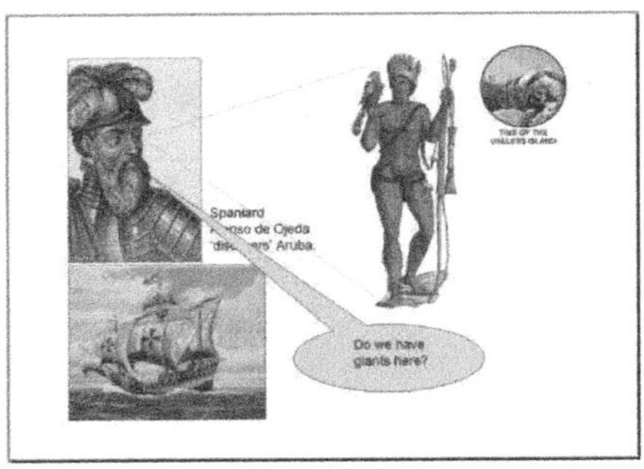
Spaniard
de Ojeda
Aruba.
Do we have
giants here?

This island is
useless!"

Spaniards import horses and goat
The whole island a large rancho.
Indians return form the continent.
handy with horses.
First Dutch visitors:
This island is not very fertile, but very barren and mostly overgrown with undergrowth; there is a large multitude of sheep and goats, with which the inhabitants take care of themselves, and also some Spanish wheat which they sow there. There are also many horses which are used by the inhabitants, because everything they do, they do on horseback even if they only go fivehundred steps from their homes to get water.

TIME OF THE USELESS ISLAND
- The Spaniards convert the Indians to catholicism.
- In the middle of the island a wooden cross is put in place: 'Santa Cruz'.

WHAT COMES TO YOUR MIND ABOUT THE TIME OF THE USELESS ISLAND?

1600
1800
TIME OF HORSES AND PIRATES

Conquest of Curaçao by the Dutch.

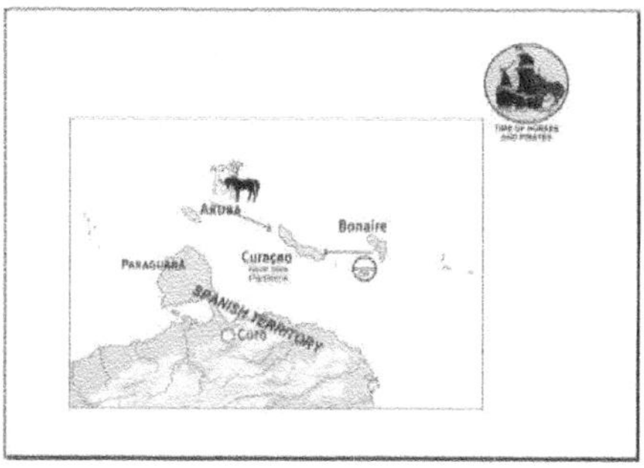
Aruba
Bonaire
Paraguana
Curaçao
SPANISH TERRITORY
Coro

Small herds of free range goats can still be seen on Aruba today...

BUCCANEERS
Aruba frequently attacked by pirates who wanted to rob horses or goats. The Netherlands could or would do little to prevent it.
Attack on a Dutch WIC ship.

Alto Vista Chapel built in 1950 at the location of the first Indian church.
The Aruban population still existed almost exclusively of Indians with a Dutch 'chief' and 15 soldiers. On Aruba almost NO slaves, on Curaçao thousands of slaves.

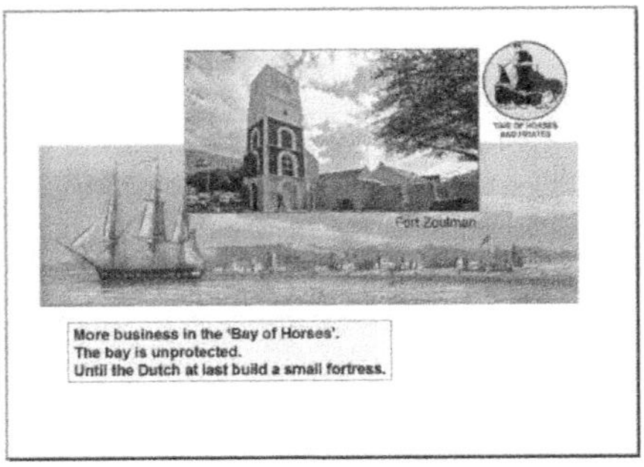
Fort Zoutman
More business in the 'Bay of Horses'.
The bay is unprotected.
Until the Dutch at last build a small fortress.

WHAT COMES TO YOUR MIND ABOUT THE TIME OF HORSES AND PIRATES?

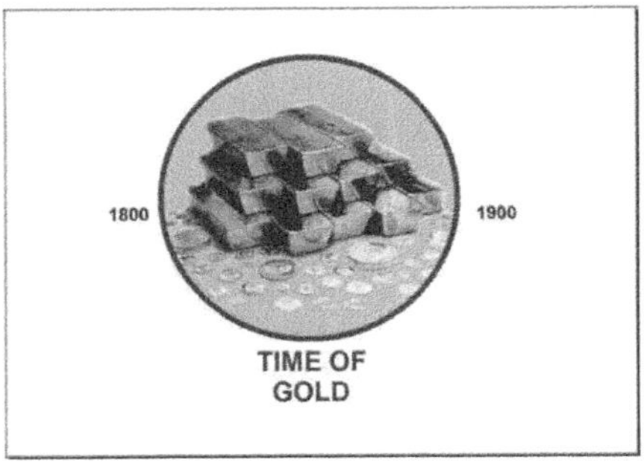
1800
1900
TIME OF GOLD

THE STORY OF WILLEM RASMIJN
If only I had a pair of shoes...

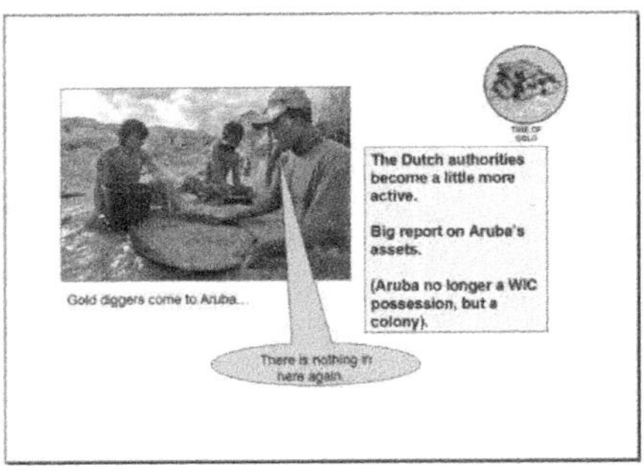
Gold diggers come to Aruba...
There is nothing in here again.
The Dutch authorities become a little more active.
Big report on Aruba's assets.
(Aruba no longer a WIC possession, but a colony).

Remains of foundries at Aruba:
Later a Gold Mining Company is founded which builds foundries.
'Tributers' everywhere on the island.

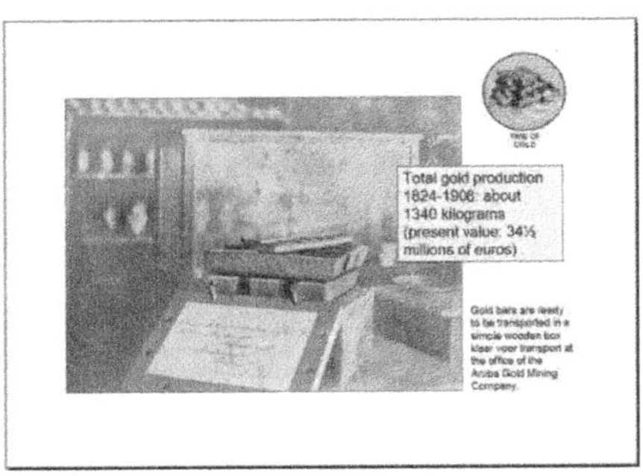
Total gold production 1824-1908: about 1340 kilograms (present value: 34½ millions of euros)
Gold bars are ready to be transported in a simple wooden box klaar voor transport at the office of the Aruba Gold Mining Company.

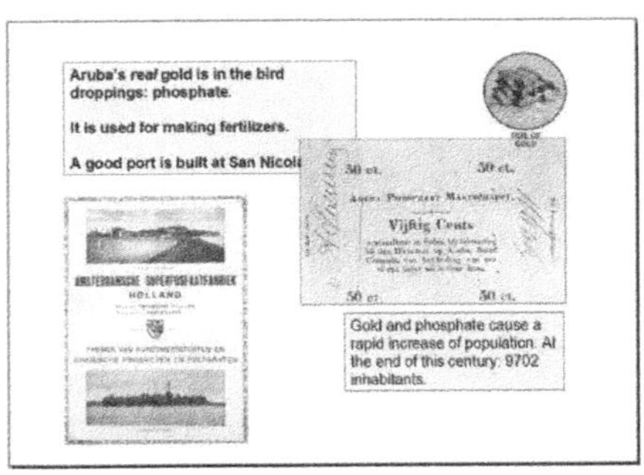
Aruba's *real* gold is in the bird droppings: phosphate.
It is used for making fertilizers.
A good port is built at San Nicol
Vijftig Cents
HOLLAND
Gold and phosphate cause a rapid increase of population. At the end of this century: 9702 inhabitants.

The port meant for export of phosphate at San Nicolas.

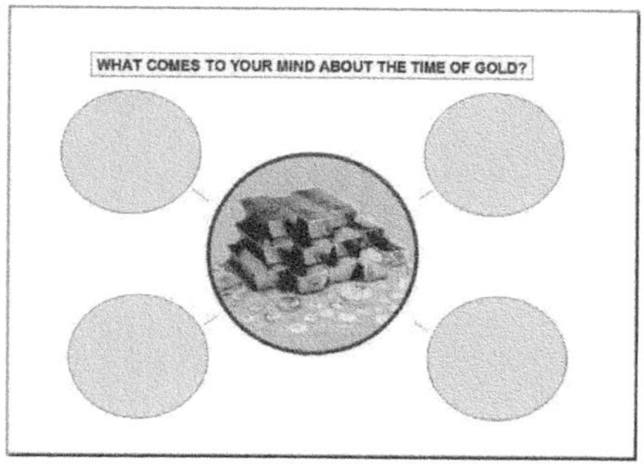
WHAT COMES TO YOUR MIND ABOUT THE TIME OF GOLD?

1900
1980
TIME OF
OIL

Governor
Frederik Quast
The visit of
'Captain' Robert
Rodgers.

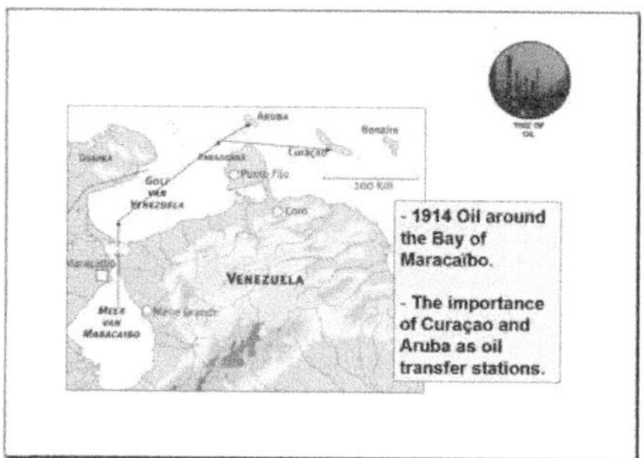
ARUBA
Bonaire
Curaçao
Punto Fijo
GOLF VAN VENEZUELA
Coro
VENEZUELA
MEER VAN MARACAIBO
- 1914 Oil around the Bay of Maracaïbo.
- The importance of Curaçao and Aruba as oil transfer stations.

De port of San Nicolas made suitbale for large sea ships.

A refinery was also built at San Nicolas.

Aruba changes.

Aruba changes.
Oranjestad main street in1925 and in 1950.

Lago Oil & Transport Co., Ltd. Aruba
The refinery becomes one of the largest in the world.

Important in the Second World War.

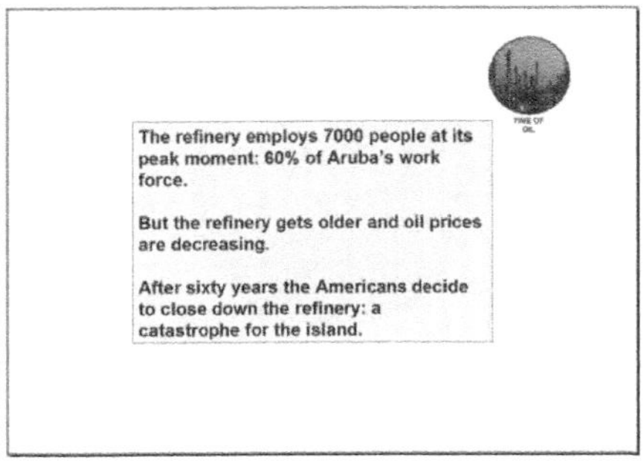
The refinery employs 7000 people at its peak moment: 60% of Aruba's work force.
But the refinery gets older and oil prices are decreasing.
After sixty years the Americans decide to close down the refinery: a catastrophe for the island.

WHAT COMES TO YOUR MIND ABOUT THE TIME OF OIL?

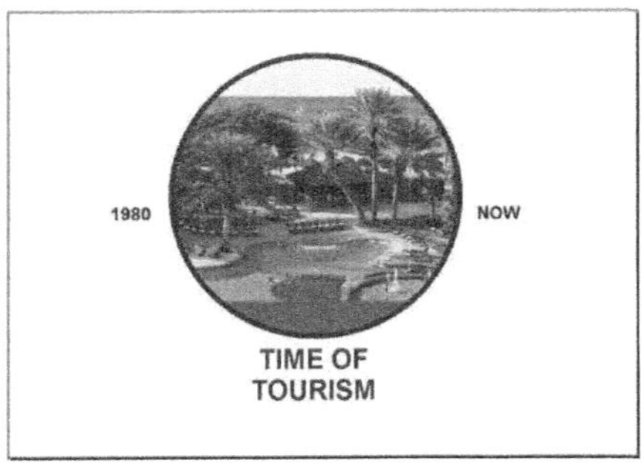
1980
NOW
TIME OF TOURISM

There once was a time when Aruba had palm beaches like in paradise...

American and Aruban businessmen make plans for the first hotel, with a famous architect.

Aruba Caribbean Hotel.

After the closing of the refinery the Aruban government did their utmost to promote tourism. This is what remains of the paradise beaches today.

More tourism.
The first somewhat rickety cruise ship and a cruise ship of today.

Aruba's government promoted tourism emphatically.
Which government?
Aruba has strived for an autonomous status within the Kingdom. Especially via the MEP party.

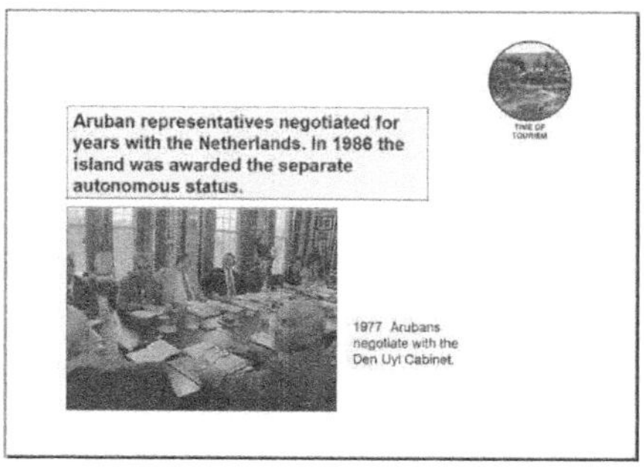
Aruban representatives negotiated for years with the Netherlands. In 1986 the island was awarded the separate autonomous status.
1977 Arubans negotiate with the Den Uyl Cabinet.

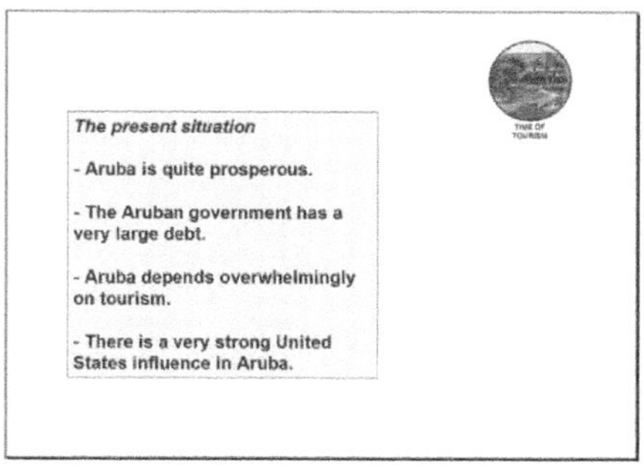
The present situation
- Aruba is quite prosperous.
- The Aruban government has a very large debt.
- Aruba depends overwhelmingly on tourism.
- There is a very strong United States influence in Aruba.

WHAT COMES TO YOUR MIND ABOUT THE TIME OF TOURISM?

WHAT COMES TO YOUR MIND ABOUT THESE SIX ERAS?
TIME OF THE INDIANS
TIME OF THE USELESS ISLAND
TIME OF HORSES AND PIRATES
TIME OF GOLD
TIME OF OIL
TIME OF TOURISM

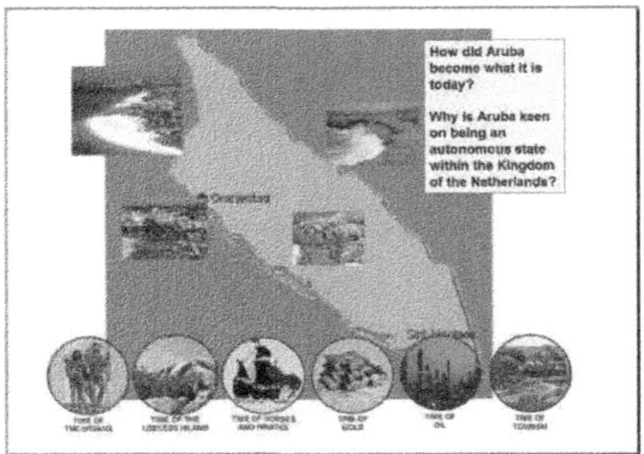
How did Aruba become what it is today?
Why is Aruba keen on being an autonomous state within the Kingdom of the Netherlands?
Oranjestad

APPENDIX B

Content Protocols for the Lessons According to the Chronological and Imaginative-Associative Approaches

Chronological

SLIDE 1: (Title)

SLIDE 2: (Question for the Lesson)

What we would like to know is: How did Aruba become what it has become? Why is it a separate country within the Kingdom of the Netherlands? And why are the Arubans keen to keep it like this? You need to look to the history of a country in order to be able to find an answer to these questions. We will be discussing the history of Aruba across six centuries. We will pause at the end of each century to note the most important

Imaginative-Associative

SLIDE 1: (Title)

SLIDE 2: (Question for the Lesson)

What we would like to know is: How did Aruba become what it has become? Why is it a separate country within the Kingdom of the Netherlands? And why are the Arubans keen to keep it like this? You need to look to the history of a country in order to be able to find an answer to these questions. We will discuss the history of Aruba in six eras. After each era we'll pause to note the most important features of the era. At the

Images of Time, pages 231–254

features of the century. At the end of the lesson you will hopefully be able to answer the questions.

SLIDE 3: (Title Slide: Fifteenth Century)

The first century we will discuss is the fifteenth century. That's the period between 1400 and 1500.

SLIDE 4: (Arowaks)

In the fifteenth century Aruba was populated by Indians. We know from excavations that Indians inhabited Aruba for centuries. The Indians inhabiting the island during the time the Europeans arrived belonged to the Arowaks, who populated a large area across South America. The Arowaks were a peaceable, friendly tribe, and completely unwarlike, both to each other and towards the Europeans. The population of Arowaks belonged to the same tribe as the inhabitants of the Paraguayan peninsula, belonging presently to Venezuela. They probably frequently sailed to and from the mainland. We know that they used canoes made from trees. However, the type of tree needed to make canoes does not grow on Aruba.

The Indians lived a simple agricultural existence (with small types of gardens where they grew some vegetables and fruits), and fished in particular. There were only a few villages on Aruba, where about a hundred to two hundred people lived. So, very few people.

SLIDE 5: (Excavations)

We have been able to learn more about the Indians of Aruba from excavations. A large number of urns like these can be viewed at the archaeological museum on the end of the lesson you will hopefully be able to answer the questions.

SLIDE 3: (Title Slide: The Time of the Indians)

The first era we're going to discuss is the Time of the Indians. That's the time before 1500.

SLIDE 4: (Archeology)

STORY—On the photograph you can see the archeologists who researched Aruba for its oldest inhabitants. They made many exciting finds. There were large ceramic vessels which had apparently been made by the Indians. The remains of skeletons were found inside those vessels. Apparently the Indians used them as urns to bury their dead. What type of people were the Indians?

SLIDE 5: (Excavation at Tanki Flip)

At the site of "Tanki Flip" (the name originates from a place where rainwater used to be collected in a "tanki" that apparently used to be owned by a certain "Flip") a large area was excavated where there used to be an Indian village. More burial urns were found at the site. A noticeable feature of the urns is the figure of a human head placed on top. Apparently the Indians wanted to keep the memory of their forebears alive in this manner. Other things belonging to the Indians were also discovered: including, for instance, the remains of houses or huts where they used to live.

SLIDE 6: (Arowaks)

Rock paintings had previously been found on Aruba. These are "abstract" drawings, that is, they do not represent any animals, people island. The remains of the dead were buried in these urns. The head on top is perhaps in memory of the dead. This urn is a beautiful copy, but this one here is one that was actually excavated—and, this one, of course, is not quite as nice. You can see there's another head on the top. Rock paintings have also been found on Aruba. These are "abstract" drawings, that is, they do not represent any animals, people or other things. We don't know what they mean.

SLIDE 6: (Alonso de Ojeda)

The Spaniard, Alonso de Ojeda, sailed past Aruba in 1499 right at the end of the fifteenth century. Alonso was one of Columbus's commanders on his second voyage. You perhaps know that Columbus "discovered" America in 1492. After that first journey he traveled more frequently back to the area around the Caribbean Sea. One of his commanders explored the southern region for him, and came across Aruba during this time. His ship must have been as large as Columbus's ship. Alonso was the first European to visit Aruba, Curaçao and Bonaire.

SLIDE 7: (Balance of the Fifteenth Century)

(Students should note important features of the fifteenth century).

SLIDE 8: (Title Slide: Sixteenth Century)

The second century we are going to discuss is the sixteenth century. That's the period of time between 1500 and 1600.

or other things. We don't know what they mean.

We do know that the Indians living on Aruba when the Europeans arrived in the area belonged to the Arowaks. The Arowaks were a peaceable, friendly tribe, and completely unwarlike, both to each other and towards the Europeans. The population of Arowaks belonged to the same tribe as the inhabitants of the Paraguayan peninsula, belonging presently to Venezuela. They probably frequently sailed to and from the mainland. We know that they used canoes made from trees. However, the type of tree needed to make canoes does not grow on Aruba.

The Indians lived a simple agricultural existence (with small types of gardens where they grew some vegetables and fruits), and fishing, in particular. There were only a few villages on Aruba, where about a hundred to two hundred people lived. So, very few people. You can see an image of a village like this here.

SLIDE 7: (Balance of the Time of the Indians)

(A number of features from the Time of the Indians are collected after a discussion with the students. The students make suggestions about what should go in the empty circles).

SLIDE 8: (Title Slide: Time of the Useless Island)

The second era we're going to discuss is the Time of the Useless Island. That's the time roughly between 1500 and 1600, approximately one hundred years.

SLIDE 9: (Useless Islands)

Soon after Alonso de Ojeda's travels more Spaniards arrived to occupy the islands. They hoped to find valuable things, preferably gold. However, after some exploration they soon realized that there was nothing to discover, no gold, and nothing else of any value. The islands were dry, rocky and barren. And so they quickly declared the islands to be "useless islands" (islas inutiles). This happened in 1513. Two years after that they were using the only useful thing on the islands: the population. They transported all the Indians away to another island in the Caribbean Sea, where gold had been found, and used them as slaves in the mines. You can probably guess what became of the Indians. No Indian ever returned to Aruba.

Sometime later the Spaniards introduced horses and goats to Aruba, to be able to do something with the island. They allowed the animals to roam around. In other words the entire island became one big "rancho," which is a Spanish word meaning a very large farm (actually more of a stretch of land), where animals can roam in herds. Indians did, of course, return from the mainland, which was so close by that it was easy to access. The Indians appeared to be particularly good at handling horses. They were skilled at catching and taming the horses. They soon became excellent riders who did everything from horseback.

SLIDE 10: (Catholicism)

There was one thing the Spaniards thought was of particular importance: their Catholic religion. Although they perhaps cared very

SLIDE 9: (Deportation of Indians)

STORY—Why does this era have such a strange name? This is the reason why. When the Spanish arrived on Aruba (which occurred quickly after Columbus's travels) they discovered there was nothing on the island they could use. And this is why they did something terrible. The only thing they could use on the island were the people who lived there. The rest of Aruba they thought was useless. That's why they transported the entire Indian population away to be used as slaves elsewhere. The Indians put up minimal resistance. After all, they weren't warrior chiefs. You might be able to guess what became of them. Not a single Indian ever returned to the island.

SLIDE 10: (Giants' Island)

The Spanish didn't immediately refer to Aruba as the "useless island" (isla inutile). They first called it "giants' island." When Alonso de Ojeda—one of Columbus's commanders on his second voyage—arrived on Aruba he thought the inhabitants were very tall. Apparently the Indians were taller than the relatively small Spaniards. "This is an island of giants (isla de los gigantes)," thought Ojeda. However, this impression did not last long. The Spanish quickly realized they had nothing to fear from the Indians.

SLIDE 11: (Useless island)

Even after careful searching the Spanish did not find anything of value on Aruba. All there was were bare rocks, cactusses and divi-divi trees. They had hoped to find valuable things, preferably gold. However, they soon realized that little for the Indians, they did believe they should become Christians, and in particular, Catholic Christians. They sent priests to the island to baptize the Indians. The first priest erected a wooden cross somewhere near the center of the island and taught the Indians to go there to recite their Rosary. We do not know whether the Indians understood much about the new religion. They slowly learned Spanish, which was the language spoken by the priests and the language in which they had to pray.

SLIDE 11: (Balance of the Sixteenth Century)

(Students should note important features of the sixteenth century).

SLIDE 12: (Title Slide: Seventeenth Century)

The third century we are going to discuss is the seventeenth century. That's the period of time between 1600 and 1700.

SLIDE 13: (Conquest)

The Dutch and English appeared in the Caribbean Sea approximately a hundred years after the Spanish had arrived. One of the causes of this was that Portugal, with an important harbor in Lisbon, was occupied by Spain at the end of the sixteenth century. The Netherlands and England were at war with Spain and were therefore no longer able to moor in Lisbon, and decided to explore the region themselves. The Dutch were particularly keen to find salt, which they needed for fishing, as salt helps preserve fish. Another reason was to inflict damage on the

there was nothing to be discovered, no gold, and nothing else of any value. The islands were dry, rocky and barren. There was nothing to be gained here. The Spanish determined officially that this was a useless island.

SLIDE 12: (Horses and goats)

The Spanish transported horses and goats to the island in order to be able to do at least something with Aruba. They allowed the animals to roam around. In other words the entire island became one big "rancho," which is a Spanish word meaning a very large farm (actually more of a stretch of land), where animals can roam in herds.

Indians did, of course, return from the mainland, which was so close by that it was easy to access. The Indians appeared to be particularly good at handling horses. They were skilled at catching and taming the horses. The first Dutch visitors arriving on Aruba noticed how attached the Indians were to their horses (Read the text on the screen).

SLIDE 13: (Catholicism)

There was one thing the Spaniards thought was of particular importance: their Catholic religion. Even though they cared very little for the Indians, they did think that they should become Christians, and Catholics in particular. They sent priests to the island to baptize the Indians. The first priest erected a wooden cross somewhere near the center of the island and taught the Indians to go there to recite their Rosary. We do not know whether the Indians understood much about the new religion. They slowly learned Spanish, which was the language

Spanish. After all, they were at war with them.

SLIDE 14: (The Three Antilles)

They discovered a beautiful place for a naval base on Curaçao very close to the coast of South America belonging to the Spanish. Curaçao has a large bay forming a natural harbor, which is why the Dutch decided to conquer the island in 1634. There was salt on Bonaire, which was also to the benefit of the Dutch. But they were less clear what to do with Aruba. They finally decided to occupy Aruba, as the Spanish might otherwise have used it as base to re-conquer Curaçao. However, the occupation should not cost too much, as Aruba had little or nothing to offer. A Dutch commander arrived with fifteen soldiers. And nobody else. The Indians, horses and goats continued to live on Aruba, as they had previously done.

This created a situation where Curaçao was an important Dutch base, and Aruba and Bonaire slightly second rate as suppliers of salt and horses.

On the map you can see how the West India Company (the Dutch trading company which had conquered the colony) had organized the three islands. Everything revolved around Curaçao. The other islands were simply suppliers, and in actual fact only occupied to prevent the Spanish doing so.

SLIDE 15: (Balance of the Seventeenth Century)

(Students have to name important features of the seventeenth century).

spoken by the priest and the language in which they had to pray.

SLIDE 14: (Balance of the Time of the Useless Island)

(A number of features about the Time of the Useless Island are gathered from the students after a discussion. The students make suggestions about what should go in the empty circles).

SLIDE 15: (Title Slide: Time of Horses and Pirates)

The third era we're going to discuss is the Time of Horses and Pirates. This is the time from roughly 1600 to 1800. This era lasted approximately two hundred years.

SLIDE 16: (Conquest)

STORY—On the picture you can see how the Dutchman Johannes van Walbeeck conquered the island of Curaçao. It looks like a large fleet of ships which violently overcame the Spanish, but the reality was quite different. The ship Groot Horn, accompanied by two yachts and two sloops, finally arrived in the bay of Curaçao after a difficult search. They had been instructed by the Amsterdam Chamber of the Dutch West India Company to conquer Curaçao and set up a naval base there for the war against Spain (the Netherlands was still involved in the Eighty Year War against Spain at that time). Curaçao had a beautiful bay which lent itself well as a natural harbor. The Spanish governor appeared under the flag of peace and asked what their intentions were: war or negotiations? They quickly realized that the Dutch meant business and rapidly fled away from the bay, so the island was conquered.

SLIDE 16: (Title Slide: Eighteenth Century)

The fourth century we are going to discuss is the eighteenth century. That's the period of time between 1700 and 1800.

SLIDE 17: (Eighteenth Century)

Aruba remained in the possession of the West India Company in this century. Very little changed on the island. There was a commander with fifteen soldiers. And there were Indians. Every now and then a Catholic priest arrived from the Spanish mainland to celebrate a mass. This had to be done in some secrecy, as it was not allowed by the Dutch commander, who was a Protestant. However, he was unable to monitor the entire island with just the few soldiers under his command. For the rest, life continued as normal. In 1750 the Indians even got their first church (up to then everything had taken place in the open air). The church was a simple building with walls made from loose stones and clay and a roof of palm leaves. The church, of course, no longer exists, but in 1950—two hundred years later—the Arubans built a chapel on the site in memory of that first church. The chapel is called "Alto Vista" or high vantage point (as you see the sea from the site). There is an old wooden Spanish cross in the church, although whether or not it's from the real first church nobody knows.

The WIC did not want any white colonists to settle on Aruba, as it saw them as competition. The Company controlled the entire island with the horses and goats and did not want any further interference. But The Dutch captured an Indian who told them that there were only six Spaniards and a hundred Indians on Curaçao. And so the Dutch came to occupy the island without a single shot being fired and started to build a fort. The Spanish were allowed to leave in peace.

SLIDE 17: (The Three Antilles)

The Spanish attempted to regain the lost island from Aruba and Bonaire. And so the Dutch decided to also occupy those two islands. Salt was discovered on Bonaire, which was a fortunate find. The Dutch needed salt for their fishing industry, as salt helps preserve fish. But they were less clear what to do with Aruba. The occupation of that island wasn't allowed to cost much, as Aruba provided very little or nothing at all. A Dutch commander arrived with fifteen soldiers. And nobody else. The Indians, horses and goats continued to live on Aruba, as they had previously done.

This created a situation where Curaçao was an important Dutch base, and Aruba and Bonaire slightly second rate as suppliers of salt and horses.

SLIDE 18: (Goats)

These days there are still herds of goats roaming around Aruba. There aren't very many, but occasionally you'll come across them. This recalls the period of time when the island was a large animal farm.

SLIDE 19: (Pirates)

This time is really the time of the "Pirates of the Caribbean." But not everything was like in the movie. The numerous islands in the Caribbean Sea were all in the hands slowly white settlers began arriving on Aruba wanting to start-up a plantation or farm of their own. Moses Maduro, a colonist from a Jewish family was the first to do so. They had to dutifully make a "pledge" to the Company similar to a serf's in the Middle Ages. They only received the ground "on loan."

Yet there were no large plantations with slaves on Aruba, in contrast to Curaçao. Which is why the population of Aruba these days is very different to that on Curaçao. On Aruba there are only people of Indian and mixed Latin American origin, while on Curaçao there are very many people whose roots lie in Africa. This is given as an explanation for the fact that the two populations wanted to remain distinct nowadays.

SLIDE 18: (Pirates)

Although the WIC did not want be involved in any way, the trade in "Bay of Horses" on Aruba gradually increased as more white settlers arrived. (The bay these days is the harbor for Oranjestad, where tourist cruise ships are moored. Its name is derived, of course, from the fact that large numbers of horses were exported from the bay).

The commander and his fifteen soldiers were unable to guarantee the safety of Aruba and the Bay of Horses. He had nothing with which to defend the island: no fort, no cannons. The WIC believed that these were too expensive. Although the Company did occasionally incur a great deal of damage caused by the pirates who brutally arrived on shore to steal the horses. There were many pirates in the Caribbean Sea in the eighteenth century. This was of different European countries, who sometimes barely governed with any real authority, such as the Dutch on Aruba. Pirates were therefore able to easily flee from island to island after making their attacks. The commander and his fifteen soldiers could do nothing to guarantee the safety of Aruba. He had nothing with which to defend the island: no fort, no cannons. The Dutch West India Company thought this would be too expensive. And so pirates regularly arrived on Aruba to raze the island to the ground and steal another cargo of horses. Sometimes this resulted in hundreds of guilders worth of damage (thousands, perhaps even tens of thousands of euros in modern day currency).

SLIDE 20: (Indians)

A few hundred Indians were living on Aruba at that time. They represented the largest proportion of the population and the Dutch commander was little more than their "chief." Every now and then a Catholic priest arrived from the Spanish mainland to celebrate mass. This had to be done in some secrecy, as it was not allowed by the Dutch commander, who was a Protestant. However, he was unable to monitor the entire island with just the few soldiers under his command. For the rest, life continued as normal. The Indians were even given their first church (up to that point services had all been held in the open air). The church was a simple building with walls made from loose stones and clay and a roof of palm leaves. The church, of course, no longer exists, but two hundred years later the Arubans built a small chapel on the site in memory of the location of caused by the fact that there were numerous islands and islets that were all governed by different European countries. It was therefore very easy to hideaway or take flight from one island to another.

The Company had calculated that defense always costs more than the damage incurred, and therefore did nothing. At the end of the eighteenth century there were approximately a thousand people on Aruba. Still very few people.

SLIDE 19: (Fort Zoutman)

Finally the levels of piracy became too much for the Company. They decided to invest some money in a very small fort and some cannons. This was Fort Zoutman (named after a Dutch admiral who had just won a battle against the English in the North Sea; the name therefore has nothing to do with Aruba).

You can see the small fort on the right that was built in 1796–1797. Low walls and a few cannons. The tower, the most eye-catching feature, actually does not belong to the fort. It was built much later as a lookout and lighthouse during the time of King William III.

SLIDE 20: (Balance of the Eighteenth Century)

(Students have to name important features of the eighteenth century).

SLIDE 21: (Title Slide: Nineteenth Century)

The fifth century we are going to discuss is the nineteenth century. That's the period time between 1800 and 1900.

the first church. The chapel is called "Alto Vista" or high vantage point (as you see the sea from the site). There is an old wooden Spanish cross in the church, although whether or not it's from the real first church, nobody knows.

In contrast to Curaçao, no plantation slaves arrived on Aruba. Which is why the population of Aruba these days in very different to that on Curaçao. On Aruba there are only people of Indian and mixed Latin American origin, while on Curaçao there are very many people whose roots lie in Africa. This is given as an explanation for the fact that the two populations want to remain distinct nowadays.

SLIDE 21: (Bay of Horses. Fort Zoutman)

The Dutch WIC did not want any white colonists to settle on Aruba, who they thought would be competitors. The Company controlled the entire island with the horses and goats, and did not want any further interference. But slowly white settlers began to arrive on Aruba wanting to start-up a plantation or farm of their own. And so the trade at the "Bay of Horses" gradually increased on Aruba. (The bay these days is the harbor for Oranjestad, where tourist cruise ships are moored. Its name is derived, of course, from the fact that large numbers of horses were exported from the bay). And of course pirates continued to land on the island to steal things. Occasionally the pirates simply took over the island. One time they even tied up the Dutch commander.

Finally the levels of piracy became too much for the Company. They decided to invest some money in a very small fort and some cannons.

SLIDE 22: (Nineteenth Century)

At the start of the nineteenth century all Dutch colonies were occupied by the English. The Company was disbanded. The English were the greatest enemies of Napoleon, and the Netherlands were in an alliance with France. At the end of the French period, the English returned most of the colonies to the Netherlands. The State now took over governing the islands.

The government ordered an extensive report to be drawn-up to discover what Aruba actually had in terms of natural resources. The governor had to complete a long list of questions. For each of the questions about what the island produced the answer was simple: "Nothing." He added that the island just produced enough for the inhabitants, but nothing more. In short: we have nothing. Still very few people lived on the island.

SLIDE 23: (Gold)

However, a few years later another message arrived from Aruba: gold had been discovered. Gold prospectors and fortune hunters streamed to the island. And, of course, the government, as you might imagine, suddenly became much more interested in Aruba. The government tried at first to stop private persons from searching for gold (they blocked off areas, but it was, of course, difficult to monitor). They wanted a company to exploit the gold (and naturally contribute a significant amount of the profit to the State). However, there appeared to be so little gold that any profitable enterprise would not be possible, and so they allowed people to freely search for gold again, until

This is Fort Zoutman (named after a Dutch admiral who had just won a battle against the English in the North Sea; the name therefore has nothing to do with Aruba). You can see the fort on the right with low walls and a few cannons. The tower, the most eye-catching feature, actually does not belong to the fort. It was built much later as a lookout and lighthouse during the time of King William III.

SLIDE 22: (Balance of the Time of Horses and Pirates)

(A number of features of the Time of the Pirates are collected during a discussion with the students. The students make suggestions about what should go in the empty circles).

SLIDE 23: (Title Slide: Time of Gold)

The fourth era we're going to discuss is the Time of Gold. This is the time from roughly 1800 to 1900. This era lasted roughly one hundred years.

SLIDE 24: (Willem Rasmijn)

STORY—There's a story that's told on Aruba of a young shepherd boy, Willem Rasmijn, who used to have to tend to his father's sheep and goat herds without any shoes. The rocky ground with sharp stones and cactus spines used to hurt his feet. He often used to ask his father for shoes. But his father always told him: "You can have shoes once you've earned them." One day he was sitting on a large boulder pulling a thorn out of his foot, when he suddenly saw something glittering in the sun. He picked it up and took it to his father. There appeared to be gold in the stone. And Willem was finally rewarded with a pair of shoes. His

1870 when a new procedure was discovered to extract more gold from the gold ore via a melting process. This led to a real mining company being set up (the Aruba Gold Mining Company), which built a number of foundries: one along the west coast (near Balashi) and one along the east coast (near Bushiribana). The remains of these can still be found on the island. However, it still remained an unprofitable enterprise if the company would dig for gold itself. "Free diggers" or "tributers" were allowed to search for rocks potentially holding gold under appalling conditions and sometimes requiring daredevil feats. They had to take great risks, as the company would only pay them if they produced rocks containing gold. The tributers couldn't or could only barely detect the gold themselves, and so you can imagine that the company was always trying to fob them off with too little money.

SLIDE 24: (Gold Production)

Not much gold was ever discovered. It took a very long time before the company could produce a reasonable gold bar. The production for an entire year was then transported to Europe in a simple wooden box. You can see the wooden box in the photo. Approximately 1340 kilograms of gold was produced on Aruba in total during the nineteenth century. That would be worth 34½ million euro at the current level of gold value. That sounds like a lot, but it's only 400,000 euro per year. Not very impressive, once you deduct the costs from it.

father continued to return to the site for days after and kept discovering more gold. He kept the place where the gold was found secret. But when he celebrated his new found wealth once too often and had too much to drink, he let slip the secret. Once it was known where the gold was to be found, everybody, of course, went to the site, including fortune-hunters from beyond Aruba. This gave rise to a gold rush on the island.

SLIDE 25: (Gold Prospectors)

Apparently the Spaniards hadn't looked carefully enough during their time on the island. But there wasn't that much gold to find on Aruba, and what was there was difficult to discover. Nevertheless the population increased considerably, because people are always attracted to gold. Levels of safety also decreased across the island, as the gold prospectors started to threaten to kill each other. The Dutch government now had to take a firm hand in governing the island. The island in the meantime no longer belonged to the Dutch WIC, which had gone bankrupt. It was now in the possession of the state. The Dutch government considered it to be a colony. The government had previously arranged for an extensive report to be drawn up about Aruba to review what they had in their possession. The governor had to answer a long list of questions. The answer to the question about what the island produced was brief: "Nothing." The island produced just enough for its inhabitants and no more. The discovery of gold changed all that. The government was very interested in the gold, because then at least

SLIDE 25: (Phosphate)

Much more important than gold was the fact that phosphate was discovered on Aruba in 1874. Phosphate is formed from bird droppings accumulating at a particular site over the centuries. The material produced contains a significant amount of fertilizer, which can be used to create artificial fertilizer. The Aruba Phosphate Company was established in 1879 when large quantities of this chemical were discovered on the south of Aruba. The Gold Mining Company was very jealous of their success. They submitted a claim in order to obtain the rights to the phosphate as well because, as they said, they had received the rights to all minerals on the island from the government. However, the courts told them that natural material produced from bird droppings was not a mineral. The Phosphate Company was therefore awarded the exploitation rights. The company was much more successful than the gold mining company. Aruba finally became profitable. However, the Arubans did not profit from this. The money earned disappeared to the Netherlands and to the colony as a whole, that is, Curaçao, which was considered to be the true center of the colony. During this time Aruba became Curaçao's cash cow—a fact that they wouldn't forget very quickly.

In the meantime the population had rapidly grown through gold fever and the phosphate industry. But there wasn't enough money in circulation for so many people, and so the Phosphate Company started to issue its own currency. the colony would be producing something.

SLIDE 26: (Gold Companies)

There was so little gold on Aruba that long-term excavations by a large company did not provide any profit for a long time. It cost more than it delivered. It was only when a new procedure was discovered to extract gold from the ore by melting it that a Gold Mining Company was set up and two foundries were built on Aruba: one along the west coast (near Balashi) and one along the east coast (near Bushiribana). The remains of these can still be found on the island. However, gold prospecting remained such an unprofitable enterprise that the gold company didn't prospect for gold itself. It allowed "free diggers," the so-called "tributers" to look for stones potentially containing gold. The tributers worked under the poorest conditions and sometimes had to perform death-defying feats. They had to take great risks, as the company would only pay them if they found stones containing gold. The tributers were barely able or not able at all to determine this for themselves, so you can understand that the company always tried to ship them off with as little money as possible.

SLIDE 27: (Gold Production)

Not much gold was ever discovered on Aruba. It took a very long time before the company could produce a reasonable gold bar. The production for an entire year was then transported to Europe in a simple wooden box. You can see the wooden box in the photo. A total of around 1340 kilograms of gold were

SLIDE 26: (Phosphate Harbor)

The only advantage to Aruba was that a major harbor was created for the phosphate ships close to San Nicolas. This is a photo taken at the phosphate harbor in San Nicolas. You can see that the loads were transported along railway tracks. A modern system.

SLIDE 27: (Balance of the Nineteenth Century)

(Students have to name important features of the nineteenth century).

SLIDE 28: (Slide Title: Twentieth Century)

The sixth and last century we're going to discuss is the twentieth century. That's the time between 1900 and 2000, or, in other words, now.

SLIDE 29: (Oil)

In 1914 oil was discovered around the Maracaibo Bay in Venezuela by a Dutch company (the Batavian Petroleum Company, a predecessor to Shell). The bay is quite shallow, so no large oil tankers can enter it. It's therefore quite handy if the oil can be transported from the bay in small ships and then pumped somewhere else into the large ocean tankers. You can see on the map that the Dutch Antilles are excellently located as an oil transfer station. It's no wonder that Shell started to use the harbor at Curaçao for this purpose. And it didn't take too long before a large oil refinery was built on Curaçao.

SLIDE 30: (Robert Rodgers)

As soon as an oil company achieves some success, its competitors naturally start to take an interest. At the start of the nineteen twenties produced on Aruba during the Gold Period. That would be worth 34½ million euro at the current level of gold value. That sounds like a lot, but it's only 400,000 euro per year. Not very impressive, once you deduct the costs from it.

SLIDE 28: (Phosphate)

Aruba's real gold appeared to be found in bird droppings. Phosphate is formed from bird droppings accumulating at a particular site over the centuries. The material produced contains a significant amount of fertilizer, which can be used to create artificial fertilizer. The Aruba Phosphate Company was established when large quantities of the product were discovered in the south of Aruba. The Gold Mining Company was very jealous of their success. They submitted a claim in order to obtain the rights to the phosphate as well because, as they said, they had received the rights to all minerals on the island from the government. However, the courts told them that a natural material produced from bird droppings was not a mineral. The Phosphate Company was therefore awarded the exploitation rights. The company was much more successful than the gold mining company. Aruba finally became profitable. However, the Arubans did not profit from this. The money earned disappeared to the Netherlands and to the colony as a whole, that is, Curaçao, which was considered to be the true center of the colony. During this time Aruba became Curaçao's cash cow—a fact that they wouldn't forget very quickly.

The population had rapidly grown during the nineteenth century through gold fever and the

American oil companies appeared on the scene, including one that would later be taken over by Standard Oil (Esso), Shell's biggest competitor. Captain Robert Rodgers, a Scottish seaman with a lot of experience in seafaring and marine commerce, was sent on behalf of this company to explore the area. Curaçao was already "occupied" by Shell, and so in 1924 he went to have a look around Aruba, where he quickly reached an agreement with the Dutch governor Quast (who to be honest had very few important matters to deal with and so was very honored by the visit) to build an oil transfer station. The harbor at San Nicolas was not really suitable for large tankers; however Rodgers wasn't very keen on having to wait for years, and so laid a large tanker to anchor along the coast as a storage facility. The small ships from Maracaibo would moor alongside this ship to unload their cargo. And once in a while large ocean ships would come along to transport the entire load. In the meantime the harbor at San Nicolas was dug out. It was ready after three years. Some oil tanks were also built on land in Aruba.

SLIDE 31: (Refinery)

Once it was apparent that there was a lot of oil available, it seemed sensible to also build a refinery at San Nicolas. That decision was made in the same year the harbor was completed, and it did not take long before the refinery was working. The refinery was called "Lago" on Aruba, which is the name of the company taken over by Esso.

The company was called Lago after the Spanish word for "lake," as the oil came out of the Lake of Maracaibo.

phosphate industry. But there wasn't enough money in circulation for so many people, and so the Phosphate Company started to issue its own currency.

SLIDE 29: (Phosphate Harbor)

The only advantage to Aruba was that a major harbor was created for the phosphate ships close to San Nicolas. This is a photo taken at the phosphate harbor in San Nicolas. You can see that the loads were transported along railway tracks. A modern system.

SLIDE 30: (Balance of the Time of Gold)

(A number of features of the Time of Gold are collected from students after a discussion. The students make suggestions about what should go in the empty circles).

SLIDE 31: (Slide Title: Time of Oil)

The fifth era we are going to discuss is the Time of Oil. This is the time from roughly 1900 to 1980. This era lasted around eighty years.

SLIDE 32: (Robert Rodgers)

STORY—You can see one of Aruba's most important visitors in the photo. This is the Scottish seaman, Captain Robert Rodgers. You can see him sitting in the middle surrounded by the most important members of his crew. He arrived in Aruba in the steamship Brion—the first time that a ship not propelled by sails had entered the Bay of Horses. This will certainly have attracted spectators. Rodgers was a very experienced seaman, adventurer and entrepreneur. Before the First World War he had been particularly involved in passenger transport, he

SLIDE 32: (Revolution on Aruba)

The arrival of the oil industry meant a revolution for Aruba. Modernization took place on the island at a high pace during the twenties and thirties. For instance, up to that point there had been no asphalt roads, so it took hours to travel the short distance between Oranjestad and San Nicolas. This was all to change rapidly: roads, water supply (water distilled from seawater, as rainfall was not enough to support all those people), electricity, schools (as people now had to work in offices), and so on. Somebody of stature was appointed as the Dutch governor, someone with experience of colonial administration in the Dutch Indies, someone unlike Quast who had only been appointed as window dressing. Wagenaar worked well with the director of the American oil company, who did a lot more than simply managing his company. He was, broadly speaking, the second governor of Aruba, taking care of everything required by the island. The main road in Aruba bears his name today: L. G. Smith Boulevard.

SLIDE 33: (The War)

The refinery became one of the largest and most significant refineries in the world. During the Second World War 80% of all allied aircraft fuel came from the Antilles. This meant that Aruba also became a target for the Germans. In 1942 they sent a submarine across the entire Atlantic Ocean specifically to attack the refinery on Aruba. The first thing the submarine did was to torpedo a few tankers originating from Maracaibo. The submarine surfaced after that to start firing at the refinery with its

had fought in the war, and he now wanted to concentrate on the kind of transport that seemed to have a future: transporting oil in ocean tankers. He had received instructions from an American oil company to investigate where Venezuelan oil could best be transferred to large oil tankers. The islands around the coast seemed to be very favorable. On Curaçao, Rodgers discovered that the suitable locations were already in possession of another oil company, namely Royal Dutch Shell. But a pilot working in Curaçao harbor alerted Rodgers to a suitable place on the neighboring island Aruba, where there was an excellent site at San Nicolas bay for a harbor, which had not been in use since the end of the phosphate industry. On Aruba the visitors lodged at the only hotel the island could boast at that time: the Pension Aruba in Oranjestad. The gentlemen were exploring the area for a site to transfer crude oil originating from Maracaibo Bay in Venezuela. Their host took the gentlemen to San Nicolas. Not such a simple journey given the route across a dismal rocky road leading from one side of the island to the other. Once in San Nicolas the host's country house was put at the disposal of the guests. The gentlemen stayed there for fourteen days, and discussions were held during that time with governor Quast, among others, about the possibility of creating a channel in San Nicolas bay, which would have to be dug across the reef. Rodgers and his crew left after a two-week stay, but they were back again a month later. A decision had been made to build a transfer and storage facility at San Nicolas for crude oil. Governor Quast, who had had virtually nothing onboard cannon, but the crew made a mistake during this and the cannon exploded. The attack was therefore halted; otherwise Aruba's fate would perhaps have turned out completely differently.

SLIDE 34: (The Nineteen Fifties)

The refinery enjoyed a growth period after the war. Around sixty percent of Arubans worked there. An American colony built up around the refinery, which was totally separate from the rest of the island (ordinary Arubans were not allowed in). This contained villas where the American employees lived and had its own high school, church, shops, and so on—even its own police. Aruba became richer and more self-sufficient.

In 1954 there was an end to the colonial status of Surinam and the Dutch Antilles. These became autonomous, independent countries within the Kingdom of the Netherlands. This meant that the queen was still the head of state of the Dutch Antilles and Surinam, and that foreign affairs and defense for those countries were carried out by the whole Kingdom (or actually by the Netherlands). For all other aspects the countries had their own government chosen by their own people. So the Antilles no longer had to obey the Dutch government.

However, the Netherlands Antilles (six islands) became one single country, of which Curaçao, of course, was the most important. The Arubans now had a sense of being governed by Curaçao instead of the Netherlands. And they didn't like that. They wanted their own government, separate from the other Antilles. But the Netherlands believed that a small island could not be a "proper" to do on the impoverished island since the end of the phosphate and gold enterprises, would have been impressed.

SLIDE 33: (Oil)

The oil came from the area around the bay of Maracaibo in Venezuela. The bay is quite shallow, so no large oil tankers can enter it. It's therefore quite handy if the oil can be transported from the bay in small ships and then pumped somewhere else into the large ocean tankers. You can see on the map that the Netherlands Antilles are excellently located as an oil transfer station.

SLIDE 34: (San Nicolas Harbor)

The harbor at St Nicolas was not really suitable for large tankers; however, Rodgers wasn't very keen on having to wait for years, and so laid a large tanker to anchor along the coast as a storage facility. The small ships from Maracaibo would moor alongside this ship to unload their cargo. Once in a while a large ocean vessel would arrive to take away the entire load. In the meantime the harbor at San Nicolas was dug out. It was ready after three years. Some oil tanks were also built on land in Aruba. In the photo you can see a large ship in the excavated harbor at San Nicolas. In the background there's a smaller "lake tanker" from Maracaibo lake.

SLIDE 35: (Refinery)

Once it was apparent that there was a lot of oil available, it seemed sensible to also build a refinery at San Nicolas. That decision was taken in the same year the harbor was completed, and it did not take long before the refinery was working.

country by itself. After all, how could you create a country of just 50,000 inhabitants? This why the Netherlands wanted the six islands to remain together.

SLIDE 35: (The Start of Tourism)

In the nineteen fifties there was still only a single hotel on Aruba, which wasn't even close to the beach. Not many tourists visited the island. But this was to change slowly. The first rickety cruise ship moored on the island in 1955. There was very little for the passengers to experience on shore, but nevertheless, it was a start. The company said it would not return to Aruba unless more entertainment was provided for the passengers.

The first real tourist hotel was built in 1959 next to the beach on Aruba, and was situated in the paradise-like environment of an almost deserted palm beach.

SLIDE 36: (MEP)

In the meantime Aruba continued to be dissatisfied with its status within the Netherlands Antilles. The more Arubans earned from oil and tourists, the more self-sufficient they became. In 1971 a former schoolteacher, Betico Croes, created a new political party, the Movimiento Electoral di Pueblo (MEP, or The People's Electoral Movement), which started to make hard-hitting campaigns for self-government of Aruba, separate from Curaçao. They called this "separate status." The MEP received a lot of support in Aruba and there were even riots where order had to be restored by police officers from Curaçao, which was just more oil on the fire for the Arubans.

It was named "Lago," the name of the oil company taken over by Esso (Shell's biggest competitor). The company was called Lago after the Spanish word for "lake," as the oil came out of the lake at Maracaibo.

SLIDE 36: (Revolution on Aruba)

The arrival of the oil industry meant a revolution for Aruba. Modernization took place on the island at a high pace during the twenties and thirties. For instance, up to that point there had been no asphalt roads, so it took hours to travel the short distance between Oranjestad and San Nicolas. This was all to change rapidly: roads, water supply (water distilled from seawater, as rainfall was not enough to support all those people), electricity, schools (as people now had to work in offices), and so on.

SLIDE 37: (Revolution on Aruba)

Arubans became richer. Shops were set up in the town and more and more cars arrived.

SLIDE 38: (Lago)

The refinery became one of the largest and most significant refineries in the world. Around sixty percent of Arubans worked there. An American colony built up around the refinery, which was totally separate from the rest of the island (ordinary Arubans were not allowed in). This contained villas where the American employees lived and had its own high school, church, shops, and so on—even its own police. Aruba became richer and more self-sufficient.

SLIDE 39: (The War)

During the Second World War, 80% of all allied aircraft fuel came from the

SLIDE 37: (Negotiations)

The Dutch government continued to believe that the separate islands could not form their own countries. However, they eventually came round to the idea of independence. For instance, Surinam received independence in 1975. Aruba continued to maintain its argument for separate status. After years of tug-of-war, an agreement was finally reached in 1983. Aruba was to receive its separate status in 1986 on condition that it would become entirely independent ten years later. It was therefore to leave the Kingdom in 1996.

SLIDE 38: (1986)

Aruba received its separate status in 1986 as agreed. However, something else happened in that year. It was a pure coincidence it happened in the same year. The Americans had come to the conclusion years ago that the refinery on Aruba needed to be thoroughly modernized. However, oil prices were very low in the nineteen eighties. The modernizations weren't affordable. They therefore decided to close the aging Lago refinery.

This was a disaster for the island. The brand new, independent government decided that they had to focus entirely on tourism in order to generate work for Aruba. They asked the Netherlands for support. The agreement about independence in ten years time suddenly sounded a bit risky. Could Aruba tackle this alone? After a few years it was agreed that Aruba would remain as a separate country within the Kingdom for good.

In the meantime the Aruban government had provided the Antilles. This meant that Aruba also became a target for the Germans. In 1942 they sent a submarine across the entire Atlantic Ocean specifically to attack the refinery on Aruba. The first thing the submarine did was to torpedo a few tankers originating from Maracaibo. The submarine surfaced after that to start firing at the refinery with its onboard cannon, but the crew made a mistake during this and the cannon exploded. The attack was therefore halted; otherwise Aruba's fate would perhaps have worked out completely differently.

SLIDE 40: (The End)

The refinery enjoyed a boom period for around twenty-five years after the war. The entire island revolved around Lago. But eventually the Americans came to the conclusion that the refinery in Aruba needed to be thoroughly modernized. However, oil prices were very low in the nineteen eighties. The modernizations weren't affordable. They therefore decided to close the aging Lago refinery. This was a disaster for the island, which was entirely dependent on the refinery.

SLIDE 41: (Balance of the Time of Oil)

(A number of features of the Oil Period are noted after discussions with the students. The students make suggestions about what should go in the empty circles).

SLIDE 42: (Title Slide: Time of Tourism)

The sixth and last era we're going to discuss is the Time of Tourism. That's the time from around 1980 onwards, so around thirty years until now.

stimulus for the construction of all kinds of tourist projects: hotels, entertainment venues, casinos, restaurants, and shopping malls. They also provided government guarantees, for which they occasionally incurred significant debts.

SLIDE 39: (The Outcome)

The outcome was that Aruba changed into one large tourist attraction, where the beautiful aspect of the palm beaches became entirely hidden behind a row of American hotels, where one cruise liner after the other moored in the harbor, where passengers were received in Oranjestad with fake Old Dutch gables in friendly colors—a huge Disneyland, in other words. Tourism generated a lot for the island, but it was also volatile. If things did not go very well in the United States for a few years then suddenly fewer tourists would arrive. Aruba became highly dependent on the United States.

SLIDE 40: (The Situation Today)

Earnings are good on Aruba. It is one of the wealthiest islands in the Caribbean Sea. However, the government took great risks in supporting all the tourist projects. Furthermore, an entire government needs to be maintained for just around a hundred thousand inhabitants: that is a large number of officials for such a small population. All in all, the government has large debts. Aruba is entirely dependent on tourism and therefore on the United States of America, which makes up the lion's share of tourists. Although Aruba has its own money, you can pay in dollars all across the

SLIDE 43: (A Paradise of Palm Beaches and the Hotel Planners)

STORY—The beaches of Aruba in the north west had been left untouched during the Time of Oil: beautiful white sand, swaying palm trees, and hardly a person in sight. Aruban constructors thought that this offered lots of opportunity. They traveled to Florida in the United States to Miami Beach with its luxury hotels in order to orientate themselves. They came to the conclusion that the beaches on Aruba were much prettier. Even prettier than those on Puerto Rico where they also went to look. They started discussions with the famous American hotel architect, Morris Lapidus, and brought him along to their island. In the photo you can see the planners busy amidst the palms. They are producing plans for a new luxury beach hotel, the Aruba Caribbean Hotel. It had to be so spectacular that it would attract the richest American tourists. A private company was created to finance the hotel. The largest shareholder was the Aruban government.

SLIDE 44: (Aruba Caribbean)

And so the first hotel was opened along the beach. As you can see, it was situated in an extensive range of palms. There were still very few tourists on the beach. The hotel guests could live in the dream of believing that this paradise island was just for them.

SLIDE 45: (Fully Built-up Coast)

But this wouldn't last long. When the oil refinery closed, other sources of income had to be explored for an island population that had rapidly expanded. The entire focus became island. Life on the island has become highly Americanized.

SLIDE 41: (Balance of the Twentieth Century)

(Students should name important features of the twentieth century).

SLIDE 42: (Balance of the Six Centuries)

(Each student should choose five significant events. These should be noted down in their time. The teacher will discuss their choices and compare them with the five he has selected on the timeline. For instance, phosphate was more important than gold in the nineteenth century).

SLIDE 43: (How did Aruba become what it is and why does it want separate status?)

The population of Aruba is different from that on Curaçao. Aruba does not have a history of slavery. The largest proportion of the population arrived in the twentieth century via migration, from (in particular) Latin America and the Caribbean region. The island has largely become what it is today through the history of the oil industry. Without this industry there would never have been as many inhabitants on the island. For instance, on Bonaire, which is a larger island than Aruba, but one that has no oil industry, there are only 13,000 inhabitants today. Because Aruba "increased in size" and became wealthy, it could start to feel equal to Curaçao and start arguing for separate status. In the nineteen eighties once a lot of people were living on the island and a reasonable standard of living had been achieved, the government on tourism. The island's government encouraged the building of as many hotels as possible. They provided government guarantees for this, which meant they were sometimes considerably in debt.

SLIDE 46: (Cruise Ships)

Cruise ships represented another type of tourism. In order to attract as many ships as possible, there needed to be entertainment available on shore for the passengers. The Aruban government also invested in that. Shopping malls, casinos, nightclubs, restaurants, and so on were built in Oranjestad. Although there had never been any Dutch gables on Aruba (as opposed to Curaçao, where there were old buildings built in the Dutch style), it was, of course, nice to present a "picture postcard" city to welcome the cruise ships. So imitation Dutch gables were constructed in friendly tropical colors. The new shopping malls were hidden behind these fake gables. Oranjestad became a kind of Disneyland.

SLIDES 47 AND 48: (Oranjestad Disneyland)

In the photos you can see how Oranjestad looks today.

SLIDE 49: (Separate Status)

The Aruban government strongly promoted tourism on the island. The island acquired its own government at the beginning of this era. The Netherlands Antilles had already acquired independence within the Kingdom during the Time of Oil. This meant that the queen was also head of state for the Netherlands Antilles and that foreign affairs and had to focus entirely on tourism to maintain the wealth of the island. This is why Aruba is the tourist attraction it is these days.

However, the closure of Lago demonstrated how vulnerable the island could be. This is why the Arubans preferred not to have full independence from the Netherlands. Once the troublesome folk from Curaçao had disappeared behind the horizon, the Arubans believed their link with the Netherlands was a good form of protection, being situated very closely to the Venezuelan coast (imagine Venezuela occupying their island—they wouldn't be able to do anything about that). So once a separate status had been achieved, Arubans thought it an excellent idea to remain a separate nation within the Kingdom of the Netherlands.

defense were organized together with (actually by) the Netherlands. For the remaining aspects of government the Antilles had their own government chosen by the people. They no longer had to obey the Dutch government.

However, the Netherlands Antilles (six islands) were one single country, of which Curaçao, of course, was the most important. The Arubans now had a sense of being governed by Curaçao instead of the Netherlands. And they didn't like that. They wanted their own government, separate from the other Antilles. But the Netherlands believed that a small island could not be a "proper" country by itself. After all, how could you create a country of just 50,000 inhabitants? This is why the Netherlands wanted the six islands to remain together. On Aruba a new political party led by the former school teacher Betico Croes, known as the Movimiento Electoral di Pueblo (MEP, or the People's Electoral Movement) fought hard for a separate government, independent from Curaçao. They called this "separate status." The MEP received a lot of support in Aruba, and there were even riots where order had to be restored by the police officers from Curaçao, which was just more oil on the fire for the Arubans.

SLIDE 50: (Negotiations)

The Dutch government believed that the small islands could not be separate countries, but they were agreeable to the idea of independence. For instance, Surinam received its independence a few years before. Aruba continued to maintain its argument for separate status. An agreement was reached after years of tug-of-war. Aruba was to receive its separate status on condition that it would be entirely independent ten years later. It was therefore to leave the Kingdom in 1996. But the latter never happened. When the oil refinery was closed, the Arubans became less self-confident. Perhaps they needed the Netherlands after all. Now that they were separate from Curaçao, they preferred to remain in the Kingdom as an independent, "one happy island." And so the situation remained.

SLIDE 51: (The Situation Today)

Earnings are good on Aruba. It is one of the wealthiest islands in the Caribbean Sea. However, the government took great risks in supporting all the tourist projects. Furthermore, an entire government needs to be maintained for just around a hundred thousand inhabitants: that is a large number of officials for such a small population. All in all, the government has large debts. Aruba is entirely dependent on tourism and therefore on the United States of America, which makes up the lion's share of tourists. Although Aruba has its own money, you can pay in dollars all across the island. Life on the island has become highly Americanized.

SLIDE 52: (Balance of the Time of Tourism)

(A number of features of the Time of Tourism will be collected after a discussion with the students. The students make suggestions about what should go in the empty circles).

SLIDE 53: (Balance of the Six Eras)

(The students should each choose a number of important features for the six time periods. The teacher will discuss their choices in class).

SLIDE 54: (How did Aruba become what it is and why does it want separate status?)

The population of Aruba is different from that on Curaçao. Aruba does not have a history of slavery. The largest proportion of the population arrived in the twentieth century via migration, from (in particular) Latin America and the Caribbean region. The island has largely become what it is today through the history of the oil industry. Without this industry there would never have been as many inhabitants on the island. For instance, on Bonaire, which is a larger island than Aruba, but one that has no oil, there are only 13,000 inhabitants today. As Aruba had become "bigger" and wealthier, it could start to feel "as good as" Curaçao, and therefore start to demand separate status. In the nineteen eighties once a lot of people were living on the island and a reasonable standard of living had been achieved, the government had to focus entirely on tourism to maintain the wealth of the island. This is why Aruba is the tourist attraction it is these days.

However, the closure of Lago demonstrated how vulnerable the island could be. This is why the Arubans preferred not to have full independence from the Netherlands. Once the troublesome folk from Curaçao had disappeared behind the horizon, the Arubans believed their link with the Netherlands was a good form of protection, being situated very closely to the Venezuelan coast (imagine Venezuela occupying their island—they wouldn't be able to do anything about that). So once a separate status had been achieved, Arubans thought it an excellent idea to remain a separate nation within the Kingdom of the Netherlands.

APPENDIX C

Summaries of the Subject Material Handed Out to the Students

Chronological History of Aruba

The Fifteenth Century (1400–1500)

During this time Aruba was inhabited by Indians. They were a peaceable and friendly people. Their victuals were produced by means of fishery and simple agriculture. Probably there were a few hundred Indians on Aruba. We know this from excavations.

1499 In this year the Spaniard Alonso de Ojeda arrived at Aruba. He was an officer on one of Columbus's ships during his second voyage to America.

Images of Time, pages 255–261

The Sixteenth Century (1500–1600)

After Alonso de Ojeda had discovered the island, it became quickly occupied by the Spanish. They were hoping to find useful things there or treasures, such as gold. But they found nothing.

1513 The Spaniards declare Aruba and the other islands officially "useless."

1515 In this year the Spaniards deported the entire population from Aruba. They were employed as slaves somewhere else.

1530 The Spaniards introduce goats and horses. Indians return to the island from the South American mainland. The Spaniards convert them to Catholicism.

The Seventeenth Century (1600–1700)

In this era English and Dutch appeared in the Caribbean Sea. They were at war with Spain and tried to thwart the Spaniards. On behalf of the Netherlands, this was done by the Dutch West India Company (WIC).

1634 The Dutch conquer Curaçao as a naval base for the war against the Spaniards.

1636 The Dutch also conquer Aruba, although they could do little with this island.

The Dutch use Aruba and Bonaire as extra resources for Curaçao. Aruba delivers meat and skins of horses and goats, as well as horses and goats themselves. Bonaire delivers salt.

On Aruba there are a few hundred Indians, a Dutch commander and some fifteen soldiers.

The Eighteenth Century (1700–1800)

Aruba remains a WIC possession. Nothing much changes, but there are a few developments:

1750 The Indians get their first real (Catholic) church.

1754 The first white settler arrives at Aruba.

On Aruba there are no plantations with slaves. The few slaves that live on the island are domestic servants. On Curaçao thousands of slaves work at the plantations.

Slowly the island becomes a little more important for trade, mainly in horses. That's why Aruba is troubled more frequently by pirates who want to steal horses. There are so few Dutch soldiers that they can do almost nothing about that. The WIC thought that building a fort was too expensive, but at last they did it.

1796/1797 Fort Zoutman is built.

The number of inhabitants of Aruba is less than two thousand.

The Ninteenth Century (1800–1900)

In this century Aruba was no longer a WIC possession, but a colony of the Netherlands.

1816 The Dutch government takes authority. It wants to know what can be found at Aruba. The answer: Nothing.

1824 Gold is discovered at Aruba. Gold diggers arrive in large numbers. But the quantities of gold that can be found are so small, that it isn't possible to start a Gold Mining Company.

1870 In this year a Gold Mining Company is founded after all. A new procedure has been invented to get the gold out of the ore by melting it. Foundries are built.

1874 Phosphate is found, which can be used to produce artificial fertilizer.

The phosphate produces far more earnings than the gold. But both extractions are halted around the time of the First World War (1914-1918). The gold is exhausted. Cheaper phosphate is found in other regions of the world. Aruba becomes poorer.

Yet almost ten thousand people live on the island now. They have been attracted by the gold and the phosphate.

The Twentieth Century (1900–2000)

1914 Oil is found in the area of the Bay of Maracaibo in Venezuela.

1924 Captain Rogers visits Aruba. He is looking for a place to start an oil transfer station. Curaçao already being occupied by Shell and Rodgers being employed by Esso, he chooses Aruba. The port of San Nicolas, which was built for the phosphate export, seems suitable.

1927 A refinery is being built on Aruba near San Nicolas.

The oil industry introduces major changes to Aruba. In one go it becomes a modern country, whereras before it had been rather backward. Thousands of immigrants come to work in the oil industry.

1942 The refinery is attacked by the Germans. It is one of the most important producers of fuel for allied airplanes in the Second World War.

1954 The Netherlands Antilles become an autonomous country within the Kingdom of the Netherlands. The Arubans are dissatisfied; they want to be their own country and do not want to resort under Curaçao.

1955 The first cruise ship moors at Aruba.

1959 The first tourist hotel at the beach is opened.

1971 Croes founds the MEP and campaigns for a 'separate' status.

1986 Aruba gets the separate status: its own government within the Kingdom of the Netherlands. The refinery is closed down. Massive unemployment.

The Aruban government decides to promote tourism. A large number of hotels is opened and a lot of entertainment for tourists and cruise ship passengers is created.

The population has increased a lot during the twentieth century: there are now around one hundred thousand inhabitants on Aruba.

Imaginative-Associative History of Aruba

Time of the Indians–until 1500

Rock paintings have been found on Aruba, and ceramic pots have been excavated. They were made by Indians, the oldest inhabitants of Aruba. The pots have images of human heads. The Indians buried their dead in them.

The Indians lived in small villages. There were perhaps a hundred or two hundred Indians on Aruba. A peaceable and friendly people.

Time of the Useless Island 1500–1600

Spanish conquerers were after discovering useful things or treasures, such as gold. But they found nothing. "A useless island," they said. Then they deported the entire population to another island to use them as slaves.

Time of Horses and Pirates 1600–1800

In this time the Dutch West India Company controlled the island. They did not know very well what to do with it. In fact, they were only interested in Curaçao. Aruba was used to deliver horses and goats. There were about one thousand inhabitants, almost all of them Indians.

There were many pirates in the Caribbean Sea. Frequently they came to Aruba to steal horses. The Dutch only had a few soldiers there. They thought it was too expensive to build a fort. They only did that at the end of this era: Fort Zoutman.

Time of Gold
1800–1900

A shepherd boy finds gold on Aruba. Gold diggers arrive in large numbers. But there is too little gold for a profitable exploitation.

Phosphate brings in much more money than gold. In fact, the phosphate is Aruba's real gold.

In this era Aruba becomes a colony of the Netherlands. The government has to become more active than the WIC used to. There are more people now (some ten thousand) and therefore more security problems.

Time of Oil
1900–1980

In Venezuela oil is discovered. The islands along the coast are very suitable as transfer stations for oil and for refining it.

Aruba changes rapidly. From poor and backward towards modern and "American." Many more people come to live there: from 10,000 towards 50,000.

A large oil refinery is built on Aruba. It is so important that it is attacked during World War II and so large that more than half of all Aruban workers are employed there.

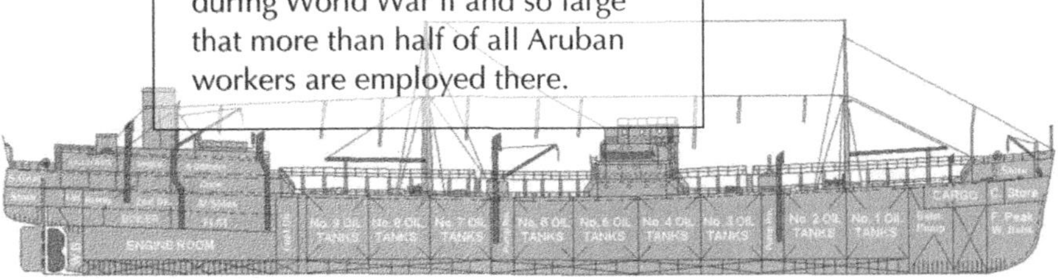

After sixty years the refinery is closed down. It is a disaster for the island.

Time of Tourism
1980–now

After the closure of the refinery the Aruban government focuses totally on tourism. The beaches are filled up with hotels.

The Aruban goverment that promotes tourism so energetically is an independent government. After a long campaign, Aruba has won the status of a separate country within the Kingdom of the Netherlands. The population increases to about a hundred thousand.

To entertain cruise ship passengers, Oranjestad is being reconstructed into a kind of Disneyland.

APPENDIX D

Questions Used for the Pre-Test and Post-Test

D.1 Prior to Pre-Test: Questions About Chronology

1. The Middle Ages lasted a thousand years and ended around 1500. How long is the period from the Middle Ages to the present day?
   A. As long as the duration of the Middle Ages.
   B. That can't be stated with any certainty.
   C. Shorter than the Middle Ages.
   D. Longer than the Middle Ages.

2. Michiel de Ruyter lived between 1607 and 1676. He captained a merchant vessel from 1622. In 1652 he became commander-in-chief of the Dutch navy. Which period of his life lasted the longest?
   A. From his youth up to his appointment as a merchant captain.
   B. The period of his appointment as a merchant captain.
   C. The period as commander-in-chief of the navy.
   D. The three periods were roughly equally long.

Images of Time, pages 263–272

3. In which century was Michiel de Ruyter alive?
   A. The fifteenth century.
   B. The sixteenth century.
   C. The seventeenth century.
   D. The eighteenth century.

4. In what sequence would you place the following events out of the life of Michiel de Ruyter? 1) He amassed great wealth in Morocco. 2) He defeated the English in the Four Days' Sea Battle. 3) He was a naughty scamp who climbed the clock tower in Flushing. 4) He was fatally wounded in a sea battle near Sicily.
   A. 1–2–3–4.
   B. 2–3–4–1.
   C. 3–1–2–4.
   D. 3–1–4–2.

5. The period of the French Revolution and Napoleon (1789–1815) can be divided into two parts: the time up to the point Napoleon came to power (1799) and the time after that. Most history books devote more pages to the French Revolution than to Napoleon's time.

   What do you conclude from that?
   A. The period of the French Revolution lasted longer than the time of Napoleon.
   B. The period of the French Revolution was shorter than the time of Napoleon.
   C. The period of the French Revolution lasted as long as the time of Napoleon.
   D. Nothing. The number of pages in a history book has nothing to do with how long a time period lasted.

D.2 Prior to Pre-Test: Questions About the History of Aruba

Questions 6 to 10 are about the history of Aruba. These are not multiple choice questions, but questions to which you have to provide an answer. It is likely that you don't know very much or don't have an answer to these questions. If that's the case, then leave the space for the answer blank and tick the box for "I don't know the answer."

6. Roughly when was Aruba conquered by the Dutch?

   Answer: ____________________

   ☐ I don't know the answer.

7. Who was in power on Aruba when the Dutch conquered the island?

   Answer: ______________________________

   ☐ I don't know the answer.

8. What minerals (e.g., coal, oil or something else you find in the ground) have been found on Aruba?

   Answer: ______________________________

   ☐ I don't know the answer.

9. For what purpose particularly was Aruba used prior to 1800 by the Dutch?

   Answer: ______________________________

   ☐ I don't know the answer.

10. What was the name of the Aruban leader who achieved a separate government for the island?

    Answer: ______________________________

    ☐ I don't know the answer.

D.3 Prior to Pre-Test: Questions About the Affinity with Aruba

11. Have you ever visited Aruba or one of the other islands of the Netherlands Antilles?

    Yes/No. *(Skip the question below if you answered "no").*

    If the answer is "yes," was that because *(put a cross next to what applies to you)*:

    ☐ You were born there

    ☐ You lived there

    ☐ You've been on holiday there

12. Do you know any people from Aruba or one of the other islands of the Netherlands Antilles?

    Yes/No *(Skip the question below if you answered "no").*

    If yes, how well do you know these people? *(Place cross where applicable)*

    ☐ Very well

    ☐ Relatively well

    ☐ A little

13. Do you think the history of the Antillean islands is an important topic for the Dutch?

    Yes/No

14. Does the history of the Antillean islands seem an interesting topic to you for a lesson?

    ☐ Yes, very interesting.

    ☐ No, not very interesting.

    ☐ I don't know/I'm not bothered.

D.4 Pre-Test and Post-Test: The Twenty-Five Events to be Located in Time

Below there are twenty-five events from the history of Aruba. You're asked to indicate when these occurred. You can answer the questions with

- a year, or
- saying "approximately around . . . (a year)," or
- "in this particular century," or
- "so many years ago," or
- "during the time of . . ."

If you don't know the answer, tick the box for "I don't know."

EVENT 1

A group of intrepid inhabitants of the South American mainland explore the empty island and decide, because there is nobody else there, to start living on the island.

When did this happen?

Answer: ______________________________

☐ I don't know.

EVENT 2

Separate fenced-off housing areas are built for Westerners arriving in Aruba to work.

When did this happen?

Answer: ______________________________

☐ I don't know.

EVENT 3

The governor orders the inhabitants of Aruba to capture a number of horses that are needed for the Dutch government on Curaçao.

When did this happen?

Answer: ______________________________

☐ I don't know.

EVENT 4

There is a great uproar in the United States as a girl on holiday in Aruba disappears and is probably murdered.

When did this happen?

Answer: ______________________________

☐ I don't know.

EVENT 5

The increasing wealth on Aruba helps the population to become more self-sufficient and consequently the people want their own government.

When did this happen?

Answer: ______________________________

☐ I don't know.

EVENT 6

A new harbor is constructed at San Nicolas for the oil tankers.

When did this happen?

Answer: ______________________________

☐ I don't know.

EVENT 7

The Holiday Inn hotel chain opens a hotel along the beaches in the northwest of Aruba.

When did this happen?

Answer: ______________________________

☐ I don't know.

EVENT 8

The governor of Curaçao visits Aruba and orders a fort to be built as a defense against the English, Spanish and pirates.

When did this happen?

Answer: ____________________

☐ I don't know.

EVENT 9

Aruba attracts a lot of adventurers from the South American mainland hoping to become rich quickly.

When did this happen?

Answer: ____________________

☐ I don't know.

EVENT 10

Governor Wagemakers of Aruba asks for more police, as there is a tense atmosphere between the Americans and the colored population on the island.

When did this happen?

Answer: ____________________

☐ I don't know.

EVENT 11

The tribal chief on the island orders a great feast to be organized in honor of the gods.

When did this happen?

Answer: ____________________

☐ I don't know.

EVENT 12

The inhabitants of Aruba are transported away to work in plantations and mines on another island.

When did this happen?

Answer: ____________________

☐ I don't know.

EVENT 13

The government of Aruba decides to allow a new shopping mall to be built in Oranjestad.

When did this happen?

Answer: ______________________________

☐ I don't know.

EVENT 14

A German submarine attacks the refinery on Aruba.

When did this happen?

Answer: ______________________________

☐ I don't know.

EVENT 15

An Indian chief—who has been christened—is brought to Aruba from the South American mainland. He has to convince the population of Aruba to convert to Christianity.

When did this happen?

Answer: ______________________________

☐ I don't know.

EVENT 16

The governor of the Netherlands Antilles is led around to inspect the new oil harbor in Aruba.

When did this happen?

Answer: ______________________________

☐ I don't know.

EVENT 17

The Dutchman Dirk van Uitgeest sails past Aruba; he doesn't dare to set foot on land, as he sees Indians on horseback.

When did this happen?

Answer: ______________________________

☐ I don't know.

EVENT 18

The Spanish pirate Balthasar Carion attacks Aruba to steal horses. He has the Dutch commander killed.

When did this happen?

Answer: ____________________

☐ I don't know.

EVENT 19

A Catholic priest from the mainland of South America is secretly brought to Aruba to celebrate a Catholic mass. Although not permitted, this still takes place.

When did this happen?

Answer: ____________________

☐ I don't know.

EVENT 20

A worker falls out of a hoist being used to extract him from a mine, and he dies within a few hours.

When did this happen?

Answer: ____________________

☐ I don't know.

EVENT 21

A mining company reports in its annual report that it is not necessary to separately list all the income from all the finds, as this amounts to very little.

When did this happen?

Answer: ____________________

☐ I don't know.

EVENT 22

There's a party to celebrate the trillionth oil barrel being refined on Aruba.

When did this happen?

Answer: ____________________

☐ I don't know.

EVENT 23

A new airport is built with a runway long enough for jet aircraft.

When did this happen?

Answer: ______________________________

☐ I don't know.

EVENT 24

Water pipelines are built on the island with water distilled from sea water.

When did this happen?

Answer: ______________________________

☐ I don't know.

EVENT 25

Commander Lodewijk Boyé—on the orders of the Dutch government—draws up an extensive report about what is to be found on the colony of Aruba.

When did this happen?

Answer: ______________________________

☐ I don't know.

D.5 Prior to Post-Test: Questions About How the Students Prepared for the Post-Test

I read through the paper that was given out at the end of the lesson

☐ Yes ☐ No

I *studied* the material in the paper that was handed out after the lesson

☐ Yes ☐ No

I didn't do anything concerning Aruba after last week's lesson

☐ Yes ☐ No

D.6 After Post-Test: Questions About the Experience of the Lesson and Teacher

1. Did you enjoy the lesson about Aruba? *Put a cross at the correct spot along the line below.*

Not at all enjoyable	Not enjoyable	Average	Quite enjoyable	Very enjoyable

2. Was the lesson about Aruba clear? *Put a cross at the correct spot along the line below.*

Not at all clear	Not clear	Average	Quite clear	Very clear

3. Did you like the teacher? *Put a cross at the correct spot along the line below.*

Didn't like at all	Didn't like	Liked somewhat	Quite liked	Liked very much

APPENDIX E

Rules for Scoring the Pre-Test and Post-Test

Item	2 points	1 point	0 points
01. A group of intrepid inhabitants of the South American mainland explore the empty island and decide, because there is nobody else there, to start living on the island.	Time of the Useless Island. Sixteenth century.		
02. Separate fenced-off housing areas are built for Westerners arriving in Aruba to work.	Time of Oil. Period 1900–1950.	Twentieth century.	
03. The governor orders the inhabitants of Aruba to capture a number of horses that are needed for the Dutch government on Curaçao.	Tine of Horses and Pirates. Period 1600–1800.		

Images of Time, pages 273–277

Item	2 points	1 point	0 points
04. There is a great uproar in the United States as a girl on holiday in Aruba disappears and is probably murdered.	Time of Tourism. A few years ago. Twenty-first century.	Twentieth century.	
05. The increasing wealth on Aruba helps the population to become more self-sufficient and consequently the people want their own government.	Time of Tourism. Period 1950–2000.	Time of Oil. Twentieth century.	
06. A new harbor is constructed at San Nicolas for the oil tankers.	Time of Oil. Period 1900–1950.	Twentieth century.	
07. The Holiday Inn hotel chain opens a hotel along the beaches in the northwest of Aruba.	Time of Tourism. Period 1980–2000.	Twentieth century.	Everything indicating a very recent development.
08. The governor of Curaçao visits Aruba and orders a fort to be built as a defense against the English, Spanish and pirates.	Time of Horses and Pirates. (End of) eighteenth century.	Seventeenth century.	
09. Aruba attracts a lot of adventurers from the South American mainland hoping to become rich quickly.	Time of Gold. Nineteenth century.		
10. Governor Wagemakers of Aruba asks for more police, as there is a tense atmosphere between the Americans and the colored population on the island.	Time of Oil. Period 1900–1950.	Twentieth century.	

Item	2 points	1 point	0 points
11. The tribal chief on the island orders a great feast to be organized in honor of the gods.	Time of the Indians. Period before 1500.		
12. The inhabitants of Aruba are transported away to work in plantations and mines on another island.	Time of the Useless Island. Sixteenth century.		
13. The government of Aruba decides to allow a new shopping mall to be built in Oranjestad.	Time of Tourism. Period 1980–2000.	Twentieth century.	Everything indicating a very recent development.
14. A German submarine attacks the refinery on Aruba.	Time of Oil. Period 1900–1950. Second World War, or 1939–1945.	Twentieth century.	
15. An Indian chief—who has been christened—is brought to Aruba from the South American mainland. He has to convince the population of Aruba to convert to Christianity.	Time of the Useless Island. Sixteenth century.		
16. The governor of the Netherlands Antilles is led around to inspect the new oil harbor in Aruba.	Time of Oil. Period 1900–1950.	Twientieth century.	Everything indicating a very recent development.
17. The Dutchman Dirk van Uitgeest sails past Aruba; he doesn't dare to set foot on land, as he sees Indians on horseback.	Time of the Useless Island. Period 1580–1640. Sixteenth century.	Time of Horses and Pirates. Seventeenth century.	

Item	2 points	1 point	0 points
18. The Spanish pirate Balthasar Carion attacks Aruba to steal horses. He has the Dutch commander killed.	Time of Horses and Pirates. Seventeenth century. Eighteenth century.		
19. A Catholic priest from the mainland of South America is secretly brought to Aruba to celebrate a Catholic mass. Although not permitted, this still takes place.	Time of Horses and Pirates. Seventeenth century. Eighteenth century.		
20. A worker falls out of a hoist being used to extract him from a mine, and he dies within a few hours.	Time of Gold. Nineteenth century.		
21. A mining company reports in its annual report that it is not necessary to separately list all the income from all the finds, as this amounts to very little.	Time of Gold. Nineteenth century. Around 1900.		
22. There's a party to celebrate the trillionth oil barrel being refined on Aruba.	Time of Oil. Period 1900–1950.	Twentieth century.	
23. A new airport is built with a runway long enough for jet aircraft.	Time of Tourism. Period 1950–2000.	Twentieth century.	Everything indicating a very recent development.
24. Water pipelines are built on the island with water distilled from sea water.	Time of Oil. Period 1900–1950.	Twentieth century.	Everything indicating a very recent development.

Item	2 points	1 point	0 points
25. Commander Lodewijk Boyé—on the orders of the Dutch government—draws up an extensive report about what is to be found on the colony of Aruba.	Time of Gold. Nineteenth century. Around 1800.		

APPENDIX F

Average Scores to All 25 Items During the Pre-Test and Post-Test

Item Number	Pre-Test			Post-Test		
	Average score	Average score Group A	Average score Group B	Average score	Average score Group A	Average score Group B
1. empty island	0.10	0.12	0.09	0.25	0.24	0.27
2. fenced-off housing	0.18	0.22	0.14	0.48	0.27	0.70
3. catch horses	0.42	0.43	0.41	1.16	1.01	1.32
4. girl missing	1.38	1.47	1.30	1.73	1.68	1.79
5. autonomy	0.42	0.51	0.33	0.90	0.88	0.91
6. oil harbor	0.37	0.44	0.30	1.07	0.80	1.34
7. Holiday Inn	0.64	0.81	0.48	1.28	1.07	1.50
8. fort built	0.48	0.55	0.41	1.00	0.87	1.13
9. adventurers	0.12	0.18	0.07	1.10	0.96	1.25
10. racial rioting	0.22	0.31	0.12	0.45	0.41	0.48
11. celebration of gods	0.46	0.54	0.37	0.98	0.84	1.13
12. inhabitants deported	0.05	0.04	0.05	0.84	0.70	0.97
13. shopping center	0.49	0.54	0.44	1.23	1.10	1.36
14. submarine	0.70	0.78	0.61	1.36	1.27	1.45
15. Christian conversion	0.07	0.09	0.05	0.48	0.45	0.50

(continued)

Images of Time, pages 279–280

Item Number	Pre-Test Average score	Pre-Test Average score Group A	Pre-Test Average score Group B	Post-Test Average score	Post-Test Average score Group A	Post-Test Average score Group B
16. governor inspects	0.33	0.39	0.27	1.02	0.75	1.30
17. Dutchman passes	0.14	0.13	0.15	0.45	0.61	0.27
18. horses stolen	0.38	0.50	0.25	1.18	1.05	1.30
19. catholic priest	0.18	0.24	0.11	0.58	0.86	0.30
20. mine worker	0.11	0.15	0.07	0.77	0.67	0.88
21. annual report	0.10	0.16	0.05	0.93	0.87	1.00
22. trillionth barrel	0.25	0.30	0.20	1.07	0.82	1.33
23. airport	0.52	0.50	0.54	1.06	0.87	1.25
24. water supply	0.39	0.50	0.29	0.68	0.67	0.69
25. report inventory	0.09	0.11	0.06	0.35	0.33	0.37

APPENDIX G

Results of the Paired t-Tests Applied to the Differences Between Item Scores on the Pre-Test and Post-Test in Group A and Group B

	Group A			Group B		
	t	df	*p*	*t*	df	*p*
Item 1	1.781	134	0.077	2.719	131	0.007
Item 2	0.726	134	0.469	6.914*	131	< 0.001
Item 3	5.402*	134	< 0.001	8.607*	131	< 0.001
Item 4	2.907	134	0.004	5.701*	131	< 0.001
Item 5	4.188*	134	< 0.001	6.348*	131	< 0.001
Item 6	4.385*	134	< 0.001	12.315*	131	< 0.001
Item 7	3.211	134	0.002	11.171*	131	< 0.001
Item 8	3.427	134	0.001	7.203*	131	< 0.001
Item 9	8.278*	134	< 0.001	13.649*	131	< 0.001
Item 10	1.554	134	0.123	4.867*	131	< 0.001
Item 11	3.296	134	0.001	8.021*	131	< 0.001
Item 12	8.151*	134	< 0.001	10.333*	131	< 0.001
Item 13	7.212*	134	< 0.001	10.529*	131	< 0.001

Images of Time, pages 281–282

	Group A			Group B		
	t	df	*p*	*t*	df	*p*
Item 14	5.686*	134	< 0.001	8.644*	131	< 0.001
Item 15	4.471*	134	< 0.001	5.979*	131	< 0.001
Item 16	4.313*	134	< 0.001	11.981*	131	< 0.001
Item 17	5.731*	134	< 0.001	1.852	131	0.066
Item 18	5.070*	134	< 0.001	11.136*	131	< 0.001
Item 19	5.982*	134	< 0.001	2.584	131	0.011
Item 20	6.220*	134	< 0.001	8.754*	131	< 0.001
Item 21	7.765*	134	< 0.001	10.214*	131	< 0.001
Item 22	7.555*	134	< 0.001	14.277*	131	< 0.001
Item 23	4.159*	134	< 0.001	7.245*	131	< 0.001
Item 24	2.428	134	0.017	5.574*	131	< 0.001
Item 25	3.102	134	0.002	4.449*	131	< 0.001

* Significant difference between pre-test and post-test ($p < 0.001$).

APPENDIX H

Significance of the Difference between Groups A and B as Summarized in Table 4.11

		t	df	*p*
All students				
Post-test across all 25 items	Total	3.620	245.041	< 0.001
	Reproduction	3.803	248.717	< 0.001
	Application	3.160	251.461	0.002
Post-test across 18 items without keywords	Total	3.060	251.002	0.002
	Reproduction	3.088	265	0.002
	Application	2.557	265	0.011
Students stating they had read or studied				
Post-test across all 25 items	Total	5.298	173	< 0.001
	Reproduction	4.920	173	< 0.001
	Application	4.875	173	< 0.001
Post-test across 18 items without keywords	Total	4.633	173	< 0.001
	Reproduction	3.900	173	< 0.001
	Application	4.237	173	< 0.001

(continued)

Images of Time, pages 283–284

		t	**df**	*p*
Students stating they had NOT read or studied				
Post-test across all 25 items	Total	0.949	89.532	0.345
	Reproduction	1.529	87.471	0.130
	Application	0.502	90	0.617
Post-test across 18 items without keywords	Total	0.533	90	0.596
	Reproduction	1.446	89.081	0.152
	Application	0.085	90	0.933

About the Author

Arie Wilschut is an associate professor in the Department of Education of the Amsterdam University of Applied Sciences (Hogeschool van Amsterdam) and a former secondary school history teacher. His work focuses on research in teaching and learning history. As secretary to the Dutch Committee on Historical and Social Education (2001), he was the designer of the ten era curriculum, which is now employed throughout Dutch history education at primary and secondary levels. With Linda Symcox, he edited *National History Standards: The Problem of the Canon and the Future of Teaching History* (Information Age Publishing, Charlotte NC, 2009). He published "History at the mercy of politicians and ideologies. Germany, England and the Netherlands in the 19th and 20th centuries" in the *Journal of Curriculum Studies* Vol. 42, Issue 5 (2010), pp. 693–723. In Dutch and German he has published numerous articles on teaching and learning history, as well as the Dutch language textbook for student teachers of history entitled *Geschiedenisdidactiek* (Teaching History), which is widely used in colleges and universities in the Netherlands and Belgium.

Images of Time, page 285

Printed and bound by CPI Group (UK) Ltd, Croydon, CR0 4YY
15/09/2026
14962368-0008